THE CRYSTAL WORKBOOK

In the same series
THE ASTROLOGY WORKBOOK
  Cordelia Mansall
THE CHINESE ASTROLOGY WORKBOOK
  Derek Walters
THE DREAMER'S WORKBOOK
  Nerys Dee
THE DREAMWORK MANUAL
  Strephon Kaplan Williams
THE ESP WORKBOOK
  Rodney Davies
THE FORTUNE-TELLER'S WORKBOOK
  Sasha Fenton
THE GRAPHOLOGY WORKBOOK
  Margaret Gullan-Whur
THE I CHING WORKBOOK
  R. L. Wing
THE MEDITATOR'S MANUAL
  Simon Court
THE NUMEROLOGY WORKBOOK
  Julia Line
THE PALMISTRY WORKBOOK
  Nathaniel Altman
THE PLAYING CARD WORKBOOK
  Joanne Leslie
THE PSYCHIC ENERGY WORKBOOK
  R. Michael Miller and Josephine M. Harper
THE REINCARNATION WORKBOOK
  J. H. Brennan
THE RITUAL MAGIC WORKBOOK
  Dolores Ashcroft-Nowicki
THE RUNIC WORKBOOK
  Tony Willis
THE TAROT WORKBOOK
  Emily Peach

THE CRYSTAL WORKBOOK

A Complete Guide to Working with Crystals

by

URSULA MARKHAM

THE AQUARIAN PRESS

First published 1988

© Ursula Markham 1988

All rights reserved. No part of this book may be reproduced or utilized in any form or by any means, electronic or mechanical, including photocopying, recording, or by any information storage and retrieval system, without permission in writing from the Publisher.

British Library Cataloguing in Publication Data

Markham, Ursula
The crystal workbook.
1. Crystals. Occult aspects
I. Title
133

ISBN 0-85030-713-9

The Aquarian Press is part of the Thorsons Publishing Group, Wellingborough, Northamptonshire, NN8 2RQ, England

Printed in Great Britain by Woolnough Bookbinding Limited, Irthlingborough, Northamptonshire

3 5 7 9 10 8 6 4

Contents

Introduction   7

Chapter
1. Meditation   9
2. Healing   18
3. The Chakras   36
4. Psychic Development   49
5. Progressing Towards Mediumship   59
6. Crystals and Astrology   64
7. Divination   73
8. Methods of Divination   91
9. Dowsing   113
10. Gem Elixirs   127
11. Growing Your Own Crystals   148

Conclusion   153
Bibliography and Further Reading   155
Suppliers   157
Index   159

To Philip and David
With my love

Introduction

It has often been said that there is nothing new in the world. Although it is true that interest in the power of crystals is relatively newly reawakened, in fact the knowledge which lies behind this 'discovery' existed in such ancient civilizations as Lemuria and Atlantis.

It was the Atlanteans who used the power and energy of quartz crystals in all aspects of their daily life. In a way still not capable of being explained in modern terms, crystals were used to light their homes and provide their means of transport. In addition — and perhaps this is more easily understood and accepted — crystals were used by the priests in all forms of healing, whether mental, physical or emotional.

Sadly, as time went on, the power and energy inherent in the crystals was misused. This in turn led to those energies becoming unstable and, eventually, causing the eruptions which completely destroyed Atlantis.

Later the Egyptians — and in particular the priesthood of that civilization — were to make use of crystal energy for power and healing. In addition they, as well as the Aztecs and Incas, used semiprecious stones as a means of divination. The seers and soothsayers of that period were able to give direction and help to those who came to consult them — a skill which vanished through the ages and has only recently reappeared.

In recent years, perhaps with the dawning of the Aquarian Age of heightened awareness, interest in crystals and gems and the powers they contain has been reawakened. All over the world individuals and groups are finding that these powers can indeed be used for the good of mankind — particularly in terms of healing and spiritual development.

We have by no means learned all that there is to learn. This Workbook can only give information on that which we already know. It is designed to help *you* discover more about crystals and gemstones and the very special energy they contain. Who knows what discovery may be unearthed by those of you who learn to work with them with sensitivity and awareness and for the general good?

'Sermons in stones,
and good in everything.'

William Shakespeare

1.

Meditation

People meditate for many and varied reasons. Some use meditation as a release from the pressures of the twentieth century world and, although no-one would claim that meditation alone can change the particular problems surrounding the individual, there is no doubt that it can make a difference to how that individual copes with those problems. Think of the busy executive, the harassed mother, the hardworking student; no amount of meditation can alter the work they have to accomplish during the course of any day, week or month, but it can help them to make use of the times between, the moments of relaxation which can be found in any life, no matter how full. Although meditation is not as widely accepted as hopefully it will be one day, none the less it is beginning to lose the connotation it once had in the minds of many as something practised only by white-robed followers of distant gurus.

Meditation is also an essential part of the development of those who wish to reach a higher spiritual plane. Perhaps they would like to become more psychically aware; perhaps they wish to develop as spiritual healers; perhaps they merely want to increase their own self-awareness in order to improve their own lives — and thereby the lives of those around them.

Whatever the reason for wanting to meditate, there can be no doubt that the process is greatly enhanced when crystals are used in conjunction with the power of the mind. It is normal, in this instance, to use either the quartz crystal or the amethyst or, in some cases, one's own birth stone; naturally the first step is to decide precisely which crystal you feel drawn to use.

The quartz crystal, when used during meditation, will focus your own inner energies and make control of the mind easier and more effective. Should your interest be more in the direction of psychic and spiritual development, perhaps you should consider using an amethyst — or even a few — as amethysts are essentially the 'spiritual crystals' and will help you to channel and enhance those abilities which are already inherent within you. When the meditation is to be used for increased inner relaxation and peace of mind and, in particular, to find a sense of direction when you might feel that you have temporarily lost your way in life, then your own birth stone might well be the most effective aid. The list which follows contains stones most commonly associated with the different astrological birth signs:

Sign	Crystal or Stone
Aries	diamond, ruby, red jasper
Taurus	sapphire, lapis lazuli
Gemini	citrine, yellow agate
Cancer	pearl, moonstone
Leo	tiger's eye, dendritic agate
Virgo	green jasper, sardonyx
Libra	sapphire, aquamarine
Scorpio	ruby, opal, red jasper
Sagittarius	topaz
Capricorn	turquoise, smokey quartz
Aquarius	amethyst
Pisces	moonstone, rose quartz

Once you have decided upon the crystal you would like to use, the next step is to acquire it. When dealing with crystals which are to be used for meditation or for healing, it is of the utmost importance that you choose your crystal for yourself. In a later chapter we will be dealing with the subject of divination and you will see that crystals to be used for that purpose can be selected for you or you can even order them by post. The reason for this is that, in the case of divination, it is the *type* of crystal which is important — whether it is an agate or a labradorite, for example — rather than the energies given off by that crystal. Once we begin to talk about crystals for meditation or healing, however, we are concentrating upon that very special blending of the energies of the actual crystal and the user and so it is essential that the user actually handles the crystal and feels attracted to it before acquiring it. At one time it was thought that such crystals should always be purchased for you and that it was wrong to part with money in order to buy it for yourself. In reality, however, all you would be doing would be exchanging the energy you have put into earning the money for the energy accumulated by the crystal in finding its way to you — and that cannot be harmful.

Of course you may be fortunate enough to live near an area where there has been recent quarrying, and in this case it may be possible to find your own crystals in the surrounding waste rock. Exciting as this may sound, it can also be extremely hazardous and certain precautions should always be taken. First and foremost, be sure to ask permission from the owner of the land before setting foot upon it. *Never* enter tunnels or sites of underground workings as this can be extremely dangerous due to the possibility of the collapse of walls or fall of rock. In addition, if the quarrying has taken place some little time earlier, take great care when walking in the area. Vertical shafts may exist which could have been covered by plant growth and you might not even be aware of their existence until too late. For the sake of their safety it is advisable never to take children onto a quarry site, however exciting the prospect may seem.

So, let us suppose that you have decided to purchase your own crystal and you have arrived at the shop or centre. How do you select one crystal from among the hundreds which surround you? This is a time to allow your intuition to guide you and you can do so in any of the following ways:

1. Close your eyes for a moment or two and when you open them you will find that one particular crystal appears to have a bright light surrounding it; this is the one for you.
2. Handle several of the crystals in turn. You will find that some of them will feel quite cold and still, but others will appear to have a positive energy of their own. When you find one with an energy (often felt as a tingling or pulsating sensation) which attunes with your own, this is your particular crystal.
3. Close your eyes and pass your hand gently over the crystals, a few inches above them. After a few moments you will become aware of an energy, almost magnetic in its feel, which will draw your hand towards a particular crystal. Still with your eyes closed, allow your hand to be drawn slowly downwards until your fingers touch the crystal which is to be yours.

Cleansing Your Crystal

In the course of reaching you, by whatever means it came, your crystal will have acquired many energies. These may be either positive or negative and will probably be a certain number of each. What you have to do, before you can use your crystal for meditation, is to remove the negative energies and leave the positive ones. It is also necessary to convert those negative energies, removed from the crystal during the cleansing process, into positive ones. It would be wrong indeed to send them haphazardly into the world when by effort of will and of concentration you can convert them into positive energies. Whichever of the following methods of cleansing you employ, therefore, be sure that you use your will at the same time, to ensure that those negative energies are converted in that way.

The following are the most commonly used methods of cleansing for personal crystals. If your intuition tells you to use a different technique, however, then you must follow that intuition. After all, the crystal is yours and will be used to enhance your own intuitive thoughts. There would be no point in denying that intuition at the very first moment.

1. Hold the crystal in running water and let the water flow over it. The ideal, of course, would be to use fresh flowing water from a stream, but running water from a tap can be equally effective. Never use hot water as this may cause the crystal to shatter. It is important to hold the crystal while it is being cleansed in this way as you will be forging the first real links between you and your new 'friend'. And, since you are hoping to work together in harmony, the sooner that bond is established between you the better. Allow the crystal to dry naturally, in the sunlight if possible as the sun has so much energy of its own that this can only serve to enhance that which is already inherent in your crystal.
2. Use a visualization method. Sit with your crystal in a quiet place of your own choosing — one where you are unlikely to be disturbed. Relax your body and calm your mind and then imagine your crystal being surrounded with a shower of pure white light. As this white light passes over the crystal be aware that it is removing all the negative energies from it, leaving only those which are pure and positive. At the same time, visualize those same negative energies being converted into positive ones so that they may be used for the good of mankind and of the world. It may be necessary to turn the crystal over and over in your hand during the course of this cleansing process — just as you would turn an item which was being washed by water — so that the white light may reach every facet of it.
3. If you either own, or have access to, a pyramid, set the crystal beneath the pyramid and leave it for at least 24 hours for the cleansing and purification process to take place.
4. Many people simply choose to ask or pray that their crystal be cleansed and that asking is sufficient provided it is done sincerely and with belief in Spirit. Remember, once again, that it is extremely important to ask that the negative energies which leave your crystal should be converted into positive energies to be used for good.

The frequency with which this cleansing routine should be repeated depends very much upon individual circumstances. Obviously it should be done before you use your crystal for meditation at all. Some people prefer to repeat the process over a period of up to three weeks, not so much to continue the cleansing as to forge the strongest possible link between themselves and their own special crystal. If this is what you prefer, then there is no harm at all in taking several days, or even weeks, to cleanse the crystal. But if, as is essential, you are going to charge your crystal after cleansing it, then a single cleansing operation would be sufficient.

There is no one in the world who is not prone to negative feelings at some time. Whether one feels depressed, unwell or unhappy with circumstances which exist at any particular time, once you have become attuned to your own crystal, you can be sure that it will pick up these feelings and absorb some of your negativity. At such times, therefore, it is advisable to recleanse the crystal — a process which will remove the negative vibrations from the crystal itself and often from its owner too.

Charging Your Crystal

Now that you have cleansed and purified your crystal, you must take time to charge it and attune to it. For this process, peace and solitude are essential.

Find a quiet place where you will not be disturbed. Sit in a comfortable chair, holding the crystal gently in your hands. Spend a moment or two studying the crystal so that you know its shape and dimensions. Run your fingers over it so that you experience the feel of it. Look into it and see it in as much detail as possible — after all in your meditations it is to become a part of you. Now close your eyes, still retaining in your mind the image of the crystal and its outward feel and appearance.

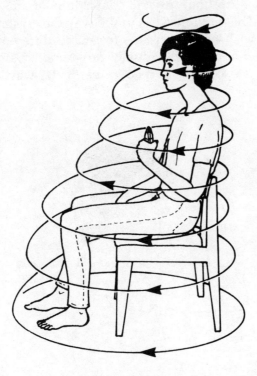

Figure 1: Holding the crystal and visualizing white light forming a spiral in a clockwise direction

Next you must spend a few moments consciously relaxing both body and mind. Start with your body and, working upwards from the feet, tense and relax each set of muscles in turn until you are free from the physical tension which we all carry around with us in the course of our everyday life. Now it is time to still your mind. That does not involve 'making your mind a blank' — something which is almost impossible for anyone other than those who spend their life in contemplation. Try instead to focus your mind upon an image which pleases you — this may be a view of the countryside, a seascape, a mountain top, the sky at sunset, or any other scene of beauty. Let your breathing become slow and regular as you begin to attune with the image and with your crystal.

As you sit there quietly, your crystal held gently in your hands, imagine a large circle of white light forming around you on the ground. Let this circle become a spiral, layer upon layer of pure white light building up around you — always in a clockwise direction — until it reaches the level above your head. Then allow that spiral to close above your head so that you and your crystal are together, safely enclosed within that pure white light. Hold that image for several moments before allowing it to fade.

The methods given here of cleansing and charging your crystal for meditation purposes will be precisely the same whether you wish to use your crystal for meditation, healing, psychic development or any other well-intentioned purpose. The difference lies in the way the crystal is programmed for your use. If you wish to use crystals for many different purposes, it is advisable to keep several, each individually programmed, and to keep them separate so that each is used only in its own particular way.

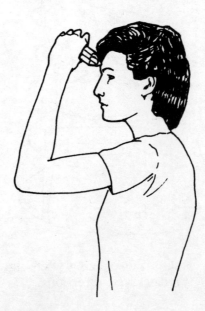

Figure 2: Holding the crystal at the level of the third eye

with practice. If you are able to set aside a particular time each day as well as a particular private place in which to practise your meditation, this too will help you to develop the ability.

Technique Two

Place your crystal on a table in front of you. Sit before it and concentrate on the crystal, focusing all your attention on it. Eventually you will feel that your eyes need to close. Do not fight this feeling but, when it occurs, allow your eyes to close naturally, trying all the while to maintain the picture of the crystal inside your mind. If the image appears to fade or your thoughts begin to wander, try to recreate the picture of the crystal in your mind. If this does not happen, then it is best to leave the meditation for that particular day since — as with the previous method — you will find that each time you try you will be able to meditate for a little longer.

Technique Three

Relax, holding your crystal in your left hand. Close your eyes so that you will not be distracted by external factors. As you breathe slowly and evenly, visualize pure white light entering your mind and your body. With practice you will actually feel the light entering every part of you as you breathe in and then, as you exhale, allow the light to leave you and spread its goodness into the atmosphere around you.

Technique Four

Place a ring of suitably charged and programmed crystals on the floor and sit on the floor in the middle of the circle. Close your eyes and allow yourself to feel the combined energy of the crystals and the interaction between them. This can lead to quite a powerful meditation and perhaps is best practised by those who already have some experience of the simpler forms.

Technique Five

Holding your crystal gently with both hands, place the diagram illustrated in Figure 3 before you and focus your attention on it. The diagram is one of the traditional focal points for crystal meditation and, although this is another form of meditation best practised by those with some experience, you will find that the basic crystal formations are represented within the design.

Technique Six

Meditation with music: holding your crystal in your left hand, and with your eyes closed, allow your mind to be led into different thought patterns by whatever gentle music you have chosen. Crystals are particularly responsive to music and you may find that, after practising this particular technique for a while, you are literally able to 'see' the music in terms of colours and shapes.

Programming

It is a known fact that crystals are able to retain patterns of thought which have been fed into them and for this reason the programming of the crystal is a significant process.

The most important factor when programming a crystal is the intention of the person whose thought processes are actually being fed into it. Knowledge as such is of lesser importance than desire and intention. It is, however, not a difficult process — although you might find that it requires some practice.

Once again, find a quiet place and spend a few moments relaxing both physically and mentally. Then, holding the crystal at the level of the third eye, concentrate on being aware of the energy link between that area and the crystal itself. Once that energy link has been established, it is sufficient to will that the crystal be of use for its particular purpose — whether healing, meditation or any other.

Meditation Techniques

There are probably as many meditation techniques as there are people wishing to meditate. The following examples are a few of the more common ones but, since we are dealing with an individual intuitive science, it is important that you use the method which feels right to you. It may be that your own intuition will lead you to select a method of your own and, if that is the case, then that technique is the one you should use.

Technique One

Because those in the West are unused to the idea of meditation, the Western mind — so used to thinking of too many things at once and rushing to fill every hour of every day with activity — must learn to be still. If this is something which does not come easily to you, this process of concentration upon breathing patterns may be of help.

Sit in a place where you can avoid being disturbed; hold your crystal in your left hand with its point uppermost. Close your eyes. Breathe in slowly, filling your lungs with air, and then release the breath again. Try to achieve a slow regular rhythm of breathing — it is often helpful to count to three while breathing in and then again while breathing out. Eventually you will reach a stage where the counting is no longer necessary and where you are able to maintain that slow even rhythm naturally and without difficulty. The advantage of this particular technique is that, if you are really concentrating upon your breathing process, it is impossible for your mind to be distracted by all the trivial little thoughts which may try to present themselves. You may well find that, when you first begin to meditate by this method, you are unable to sustain it for very long before your mind begins to wander and to become engulfed in day-to-day problems. It is better not to force yourself to continue but to accept the fact that the meditative state does become easier to maintain

Technique Seven

Group Meditation: the first priority in such a case is to discuss with other members of the group the precise intention of the meditation and ensure that there is a single purpose so that no conflict arises. There are several ways of conducting a group meditation. Some groups will place a single very large crystal in the centre of the room and then sit round it; some will prefer to have several crystals in the centre; others will each bring their own personal crystal with them. Whichever method is chosen, assuming that the group is coming together for the universal good, then very high levels of meditation can be reached.

Whichever form of meditation you have chosen for yourself, its true value will only become apparent if you make it part of your daily routine. It is not the length of time spent on each occasion in meditation which is as important as the regularity of it. Whatever your long-term intention — whether you wish to increase your psychic ability, help people individually or on a greater scale, or become more attuned to your higher self — meditation is the first and essential step. Time spent increasing your awareness in this way is time well spent indeed.

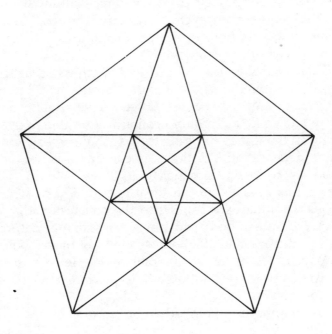

Figure 3: Traditional focal point for crystal meditation

2.

Healing

For as long as man has existed there have been healers. Methods employed may have varied throughout the ages and among different civilizations but perhaps one of the oldest of all forms of healing was with crystals. The ancient priests in Lemuria and Atlantis recognized the power of crystals in the healing process. The Egyptians, the Aztecs, the Incas and the Aborigines all used crystals in some way when helping to alleviate the problems of the sick — whether those problems were physical, mental, emotional or spiritual.

Groups of people exist today who use the energy of crystals to promote healing of the *world* — as opposed to the individual — and we can only pray for their success. This chapter, however, concentrates on healing the individual person; after all, if all people were healed, in mind as well as in body, the world would be healed at the same time.

What does *healing* actually mean? It involves treating not just the symptom, which may be obvious, but the whole person. This may necessitate treating the physical body, the aura or the chakras. Of course some illnesses are actually karmic in nature and so no amount of healing will be able to overcome them. This does not mean that those suffering from such illnesses are not entitled to love, care and understanding, nor does it mean that it is not the duty of any healer to do his best to help to conquer those illnesses in any way he can.

It would seem that crystals are an ideal energy source where healing is concerned. Scientists today are coming to the conclusion that much of the matter of which the human body consists is crystalline in nature.

In his book 'Cosmic Crystals', Ra Bonewitz talks of those bodies, other than the physical body, which make up the total being. He calls these the 'subtle bodies' and says that they are composed of levels and states of energy. These subtle bodies, too, can be treated with crystals.

Ra also makes an important point when he says:

'It is important to remember that healing does not *begin* in the physical body but in the subtle bodies of which the physical body is only a reflection. And, since crystals are capable of dealing with energies from the lowest, such as electricity, to the highest, such as subtle spiritual energies, it becomes possible to treat a number of bodies simultaneously.'

At the end of this chapter you will find a comprehensive list of the healing properties of various crystals and gems, but actual healing involves far more than just consulting such a list to find out which type of stone to use and where on the body to place it. Let us not underestimate the importance here of the intuition of the healer when sensing what it is that the patient actually needs. Equally, the importance of the will and desire of the healer to be of help must be stressed. Healing is certainly not merely a mechanical process but a true spiritual link between healer and patient — and this link exists whether the patient is present or whether healing takes place in his absence.

Whilst you will find in this chapter various techniques for diagnosis and healing with crystals, in no way is it being suggested that a patient should abandon any orthodox medical treatment he may be receiving — or indeed should fail to seek traditional medical assistance. Crystal healing and orthodox medicine can go hand in hand, with the former adding extra power and often speed to the benefits of the latter.

If you have decided to use crystals to assist you in healing yourself or others, then naturally you have to acquire, cleanse and charge the crystals for this use. It is not advisable to use a crystal which is also employed to help you in any other way — for meditation for example — as there should be different crystals for different purposes. The method of cleansing and charging the crystal will be exactly the same as that set out in Chapter 1. Once again, if a crystal is to be used for healing, it should ideally be one which you choose for yourself and not one which is selected for you. You and your crystal are to work together as a team and, therefore, you will have to be compatible from the very beginning. Just like any team of people who start off on the right footing, the more you work together, the closer and more compatible you will become.

A healing crystal should never be left lying around to collect dust. Always keep it protected, either by wrapping it in soft tissue paper or even by having a small pouch of soft material in which to keep it. If it is a personal crystal or stone which is to be used for self-healing, perhaps you could have it mounted on a chain and wear it around your neck.

Because you will be using your crystal to draw out negative energies during the course of the healing process, it is absolutely essential that the crystal itself should be recleansed each time that you use it. Whichever method you decide to use to do this, it will take no more than a few minutes of your time, but it is imperative that this cleansing becomes part of the routine or you may end up doing more harm than good.

Diagnosis

Sometimes a patient will know precisely what is wrong with him — either because the problem is obvious in nature or because there has been a medical diagnosis which there is no reason to doubt. In many cases the patient *thinks* he knows what is wrong with him — which is not the same thing at all. He may well be recognizing a symptom when the real problem is much more deep-rooted and obscure. He may even have allowed his mind and body to *produce* a symptom so that he does not have to bring himself to face a problem of which he is subconsciously very afraid. It is not really important whether the patient does know the real cause of his problem or not; any healer, whether using crystals or any other means, will want to complete his own diagnosis — although part of this will naturally consist of questioning the patient as to his symptoms and also any comment and diagnosis put forward by his own doctor.

There are various methods of diagnosis, some of which are set out below. Remember, however, that your own intuition is the greatest tool you have in

Name of patient:

Diagnostic technique(s) used:

1.

2.

3.

4.

Results:

1.

2.

3.

4.

Results of diagnosis by patient's doctor or other healer:

this regard, and if the method you use does not appear in the list, that does not mean that it is not the right method for you.

Experimenting being the greatest teacher of all, it is up to you to experiment with the various diagnostic techniques set out here to see which one suits you best. Try each one several times until you find the one which most easily enables you to attune with the pain being suffered by the patient. Obviously it is not being suggested here that you set out to be the sole helper of a person in pain or distress until you have sufficient knowledge and experience. Perhaps you could ask a friend or a member of your family — one who is already receiving treatment from a doctor, a healer, or both — to allow you to practise your diagnostic abilities on them. Keep a record of the results you get and see how these tally with the facts already known, and you will soon be shown which is the best method for you to use. A good idea would be to keep a simple card or page of a notebook for each experiment you do. Opposite is a typical example of such a record card.

1. Experiencing the pain

The healer should sit facing the patient and each should take time to relax both physically and mentally. This is not a process which should be rushed. Both healer and patient should have their eyes closed so that nothing may be allowed to interfere with the guidance and help being given to the healer. Eventually the healer should experience discomfort in the area of the body which corresponds to the area of the body of the patient which is in need of healing. This feeling should be no more than discomfort and is there not to cause pain to the healer but to serve as an indication of the part of the patient towards which healing should be directed. If this is the method you choose to use, do not forget — once you have experienced and identified the discomfort — to ask those who guide you to remove the feeling from you as the problem has now been diagnosed and recognized as being one which belongs to the patient and not to you. If you are using a crystal to aid you in diagnosis by this method, then hold your personal healing crystal lightly in both hands, pointing it towards the patient during the meditation.

2. Diagnosis from the aura

There are many healers who are able to see the aura but this is by no means true of all. Whether or not you can see it, however, does not affect whether or not you are able to diagnose from it. There are many things which one is able to 'know' spiritually without seeing them. And a known fact is no less valid because vision does not play a part in that knowledge. After all, as you read this now, do you have to look at your feet to know which shoes you are wearing? Similarly you do not have to be able to see the aura in order to realize that it is there and even to be able to feel it.

Try the following experiments to see if any or all of them feel as though they are right for you:

Sit face to face with the patient, each of you closing your eyes, and spend a few moments in quiet meditation. Open your eyes and look straight at the patient and see whether you are able to make out the aura around his head and body. If you do see it, do you see an outline of light or do you see different colours? If what you see is a natural-coloured light around the patient, is that line of light whole and unbroken or are you aware of any gaps? These gaps, should you be able to see them, will indicate to you the area of distress. If you perceive the aura as being made up of areas of different colour, take note of which colour is where. There is no set rule as to which colours indicate soundness and which indicate troubled areas — there are as many 'correct' lists of meanings of colours in the aura as there are aura readers. It is up to you to find, by experiment, which colours indicate good health to you and which show that there are problems. In this case, of course, your own record cards will be of even more importance as it is only by this sort of experiment with a number of cases that you can glean the necessary information.

Perhaps your way of diagnosis from the aura will be by feeling the aura iteself rather than by observation. If this is the case then stand close to the patient and run your hands over the aura, working from the head to the feet. You should be able to feel a warmth where the aura is whole. If you come to an area where you become aware of a chill (and you will certainly notice the contrast) then this will indicate to you the area where the problem is to be found.

Types of Crystal Healing

Whichever method of crystal healing you choose — whether it is one of the following, or one to which you feel instinctively drawn, there are a few preliminaries which should always be observed.

1. Always use a crystal which is personal to you and which you have cleansed and charged in the proper way.
2. In the quiet moment of meditation which should always occur before you attempt to heal yourself or another, remember to ask for protection for both you and your chosen crystal. You do not want to absorb the negative energies which you are helping to remove from the patient.
3. Be aware that many people, even though they may be willing to accept healing, are a little nervous of what is to them 'the unknown'. Time taken to reassure the patient and to discuss with him what you are going to do is time well spent. It is also as well to point out to him that healing — with or without crystals — is not the same as using a magic wand; there will be no 'instant cure', although there may be some immediate relief. Remember too that it is not your place to make promises of complete cures or to attempt to persuade anyone to give up any other form of treatment they may be having. You and your crystal are hoping to improve the situation and to help that person become as well and fit as he is meant to be.

The stones or crystals you use in your healing sessions will depend very much upon the type of healing you choose to do. Naturally you will use your own special healing crystal but, in addition, you may wish other crystals to be employed either by you, the patient, or both of you. The list of appropriate stones and crystals at the end of this chapter is the most generally accepted one, although it must be pointed out that there is no one comprehensive list upon which all experts agree without dissension.

Healing the aura

Hold a crystal in your left hand (or, if desired, hold two compatible crystals, one in each hand) and pass the crystals gently over that part of the aura where it seems to be broken. As you do so, visualize the energy inherent in the crystal sealing the aura so that it becomes whole again. This particular process may need to be repeated at regular intervals over a period of time.

To give you an example of the efficacy of this type of aura healing, consider the story of Alice.

Alice is a lady in her late sixties. About two years ago, during a bad winter, she fell on her garden path and hurt her right arm and shoulder, although fortunately no bones were broken. Although she suffered considerable pain and discomfort, within a few weeks the bruises disappeared and Alice was able to use her hand and arm again quite normally. There still remained a problem with her shoulder, however. Although X-rays and examination revealed that nothing had been dislocated, none the less Alice was unable to lift her right arm any further than the level of her shoulder. It wasn't that she felt any particular pain — the arm simply would not go any higher. This problem caused her quite a lot of inconvenience — she could not use her right arm to take cups or plates from the cupboard in her kitchen for example, and she had to hold her hairbrush in her left hand, although she was naturally right-handed.

Alice was not a particular believer in healing of any sort, other than that which could be received from her doctor but, when a friend who also happened to be a healer asked if he could help, she felt that she had nothing to lose and so she agreed. Twice a week the healer visited Alice in her home and, using his quartz crystal, tried to heal the aura round her right shoulder. For about three weeks there was no result — except that Alice stated that her shoulder was getting warmer during the course of the healing treatment itself. After three weeks, however, two huge purple bruises appeared which covered the whole of the afflicted shoulder. Once these bruises had faded away, Alice found that she was once again able to lift her right arm as high as she had done before the accident.

Contact healing

For this the patient should relax, preferably lying on a bed or a couch. Crystals of various types can be laid on the appropriate parts of the body and then both patient and healer should meditate and concentrate on the positive energies from those crystals flowing into the afflicted areas of the body itself. If desired, the healer may also use his own personal crystal to enhance this process but, whatever method is used, it is essential once again that *all* the stones and crystals be cleansed before and after the treatment.

Figure 4: Crystals placed on various parts of the body to accelerate healing

Another method of contact healing is for the healer to hold his own healing crystal in his left hand and rest his right hand lightly on the part of the patient's body which is in need of treatment.

Linking of energies

Patient and healer should each hold a crystal. It is of the utmost importance that these crystals are completely compatible. In some cases a crystal is cut and each person holds one half of the cut crystal. If you are to use this technique, do ensure that the crystal is recleansed and recharged after having been cut and before it is used for healing. Both patient and healer should sit, facing one another, in quiet meditation, during which time they should concentrate on visualizing the combined energies of their crystals being directed — often in the form of light or heat — towards the afflicted area of the patient's body.

Wearing or carrying the crystal

A stone or crystal should be chosen, either from the list at the end of this chapter or intuitively by the healer. This stone may be worn on a chain around the neck or carried on the person, perhaps in the pocket. If it is to be carried, however, it is advisable to wrap the crystal in a piece of silk cloth to prevent it being chafed by other items and also to ensure that, having been charged, it does not attract unwanted negativity from any other source. Some people like to sleep with their healing crystal beneath their pillow and there is certainly no harm in doing so, although of course there is not then any actual contact between the crystal and its owner.

Use of pressure points

For those with a knowledge of acupressure, it has been found that using a crystal on the appropriate acupressure points has made this treatment even more effective. If you are to use a crystal for this purpose, however, it is essential that the point of the crystal be rounded off so that it cannot pierce the skin of the patient. Even if you are employing a crystal which has already been prepared for use in healing by being properly cleansed and charged, its energies will have been damaged and altered by the cutting and smoothing process and it is extremely important that the cleansing and charging is repeated before it is placed on the pressure points of the patient's body.

Important
In some Eastern countries one can still find those who will grind the crystals into a powder and then take them internally. Even if you hear reports of this treatment being successful (and there is little or no positive verification of this being so) it is something you must *never* do. Taking crystals internally in this form can prove dangerous or even fatal!

Self-healing

Even healers need to be healed from time to time! Any of the methods already given for healing may be employed upon yourself with success. You could use your own healing crystal, you could select an appropriate stone from the list or you could merely select the type of crystal or stone which you intuitively feel is the right one for you. If you find this difficult, perhaps you could use a pendulum to help you, following the technique set out in the chapter on the pendulum which is to follow later in this book.

If you decide to choose a stone from the list, it is still best to make the final selection in person. Suppose, for example, that you have decided that the peridot is the stone which would be the right one for you to use. Ideally you should make your choice from a selection of peridots — perhaps at a shop or a centre which specializes in selling crystals. Place the peridots on a cloth in front of you, close your eyes and, when you open them, select the stone to which your eye is immediately drawn. Alternatively, you could close your eyes and hold your hand a few inches above the selection of peridots and wait until you feel either a tingling or a magnetic sensation from one of them. (Once you are adept at using a pendulum, this would be yet another way of making your final selection).

Emotional healing

Those who suffer from emotional distress are just as much in need of help as those who suffer from some physical pain or disease. One only has to consider how much of any doctor's or therapist's time is spent trying to help those with deep emotional problems to realize how large a proportion of the population we are talking about. In some ways an emotional problem is more difficult to help than a purely physical one as firstly it is harder to diagnose, secondly the patient has often allowed his emotional state to deteriorate considerably before even realizing that there was anything wrong, and thirdly some people still feel that there is some sort of stigma or shame in admitting that they need help emotionally.

Crystals have been found by some healers to be particularly effective in helping to overcome emotional problems of all sorts. It is certainly an area worthy of experimentation. It is felt that rose quartz in particular has a calming effect upon the patient, particularly when used in conjunction with relaxation and positive visualization techniques. The patient should either sit in a chair or lie on a bed or couch with a piece of rose quartz in each of his upturned hands while the healer does his work.

If a patient has a tendency to suffer anxiety attacks, it is thought that wearing a rose quartz around his neck or on his person can help to prevent these or at least to lessen their effects.

Naturally, in the case of patients suffering from any form of emotional illness, it is not being suggested here that using crystals should be substituted for any other treatment which he may be undergoing or which may be prescribed. As with physical healing, it is more a case of enhancing and accelerating other forms of treatment, whether orthodox or complementary.

There is even a school of thought which believes that crystals could be useful in the control of epilepsy. It is said that during an epileptic fit the subtle bodies are out of alignment and that crystals could help to combat this. Perhaps at some time in the future a way may be found for crystals to help in other cases of degeneration of the brain cells — either to heal actual brain tissue or to help the healthy brain cells take over the work of those which are less healthy or which have ceased to function altogether.

Spiritual healing

There will always be those who feel themselves spiritually vulnerable. Some will claim to be under psychic attack either from a known or an unknown source. It does not matter, in effect, whether this psychic attack or spiritual vulnerability is real or imagined. The terror often felt by the sufferer is genuine, even if the cause of the fear is only in his own mind. In either case it is possible to teach the sufferer one of the traditional forms of psychic self-protection and this can only be enhanced by the use of crystals. As the amethyst is known as the 'spiritual stone', it is often to be used in cases of this kind.

The person under attack should hold a large piece of amethyst in both hands — or even a piece in each hand. Closing his eyes he should imagine himself surrounded by a pure white spiritual light. This process should be repeated daily for about three weeks or until the sufferer feels safe once again. The amethyst may then be worn or carried to continue the protective process.

Although it is quite possible for someone to perform this type of spiritual healing on himself, it is often better, especially during the early days of the treatment, for him to have the additional assistance of a healer. The fear and vulnerability felt by the sufferer, by their very nature, would be sufficient to prevent successful visualization if there were no outside help. Once the patient reaches the stage where he feels that he can continue the process unaided, then the healer will recognize that the improvement has begun and that the sufferer is beginning to have more confidence in his own spiritual strength.

Group healing

Any of the healing processes detailed here may be performed by an individual or by a group of people coming together for that specific purpose. It is important, of course, that the members of the group feel drawn together for the common purpose and that there is no disharmony within the group itself. As the crystals used should also be in complete harmony, it is often beneficial to have a preliminary session where each member of the group can bring his own crystal and where the group as a whole may be involved in the communal charging of these crystals for healing purposes. Once this has been done, any members of the group — or, indeed, the entire group — may join together to add even more power to the healing that they are giving.

Absent healing

Healing of any sort does not have to be given in the physical presence of the patient. Absent healing is extremely effective. For those who think that healing is merely a question of the faith of the patient and the belief that healing will take place (some would call it self-delusion), there are innumerable instances of the success of the absent healing when the patient has not even known that he was in receipt of it. And, of course, we must remember the animals who have been helped by healing, whether present or absent, with or without crystals. These creatures have not been capable of the 'blind faith' which it is alleged causes many so-called 'cures'. But they have been helped none the less.

Absent healing may, of course, be conducted by an individual or by a group of healers coming together for the purpose of sending helpful vibrations to others. Whichever is the case, crystals are felt to enhance and improve the process. If there are to be regular sessions of absent healing by a group of people with a desire to use crystals in this way, it is often advisable to have several large crystals which, having been suitably charged, may be placed in the centre of the room, the members of the group seating themselves in a circle around them. Each member of the group should name those whom he feels are in need of healing and, after several minutes of quiet meditation, the members of the group should visualize the healing energies of the crystals being sent in the form of light and heat towards those in need. Because a tremendous amount of healing energy will be created during a session, it is as well before finally breaking the circle to ask that this energy be allowed to spread outwards to give help to the world in general.

Some absent healers like to give their patients a specific time at which they should meditate while holding a crystal. If healer and patient are working together at the same time, even if they are separated by great distances, the energy generated by the healing and the crystals can only be increased.

So now it is up to you to experiment, to try the various methods of crystal healing and to make a note of the results achieved and the benefit felt. Take your time over this. Any healing process is best practised over a period of time. But, no matter which method you choose, you will be doing your best to help your fellow man — and thereby mankind in general.

Healing Properties Attributed To Various Stones

There are many lists of crystals and stones and the healing properties attributed to each. In some cases these lists and the details contained in them do not tally. The list which follows indicates the healing powers most usually thought to be possessed by those crystals and stones which are quite easily obtainable. Remember, however, that if you feel intuitively that a crystal other than the one listed would be most appropriate in a particular situation, you should follow that intuition. Remember also that what follows is not intended to be a 'prescription for cures' but merely to form a basis for research and experiment and that this healing should go hand in hand with any other treatment already being received by the patient.

Agate The agate is said to improve natural vitality and energy and to increase the confidence of the user. It is believed to be of particular benefit to athletes or to those taking examinations — or anyone else who has to call upon instant bursts of energy — whether mental or physical.

Amazonite Generally used to help soothe the nervous system and give some relief to those who are suffering from emotional disturbances.

Amber Although it may look like one, amber is not really a stone but the fossilized resin of trees. It is claimed that amber is used to best effect by those who suffer from throat infections, having bronchial disorders or are prone to asthma or convulsions.

Amethyst It is not for nothing that the amethyst is known as the spiritual stone. It has the reputation of helping those who wish to develop psychically and of giving protection to those who feel that they may be under some sort of psychic attack. It is said that sleeping with a piece of amethyst beneath one's pillow promotes

Healing

intuitive dreams and inspired thought. Many healers also consider it useful in the relief of insomnia and to bring serenity in times of grief.

Aquamarine The aquamarine is supposed to be most useful when dealing with problems of the eyes, the liver, the throat and the stomach. It is also reputed to promote clear and logical thinking and for that reason is often carried as a 'good luck charm' by those who are taking examinations or being interviewed for a job.

Aventurine It is said that aventurine is useful in relieving migraine and in soothing the eyes. A traditional method of using it is to leave a piece of aventurine in water overnight and to use that water the following day to bathe the eyes. Aventurine water — made in the same way — can also be used for bathing irritations of the skin, with reputedly good results.

Azurite The azurite was recommended in the writings of Edgar Cayce as an aid to psychic development. It is a powerful stone and is said to have been used as a general healing stone in the lost continent of Atlantis.

Beryl It is claimed that the chief benefits of the beryl are to bring relief to those suffering from complaints of the throat and of the liver.

Bloodstone It is believed that the bloodstone will help to overcome depression and melancholia — especially if it is worn by the sufferer. It is also said to help those who suffer from psychosomatic illnesses and pains which have an emotional rather than a physical cause.

Carnelian The carnelian has the reputation of staunching the flow of blood from an open wound. If worn in a pouch around the neck by a woman during menstrual periods, it is said to help to ease the stomach cramps often felt at that time. It can also be used instead of the traditional remedy of a key placed against the back of the neck of a person who is suffering from a nosebleed.

Cat's eye It is claimed that the cat's eye will be of great benefit to those who suffer from acne, eczema and other eruptions of the skin. It also has the reputation of being a 'good luck charm' and of bringing good fortune to the wearer — but whether this has any basis in fact is doubtful.

Chalcedony This stone is said to reduce irritability in the wearer and to increase feelings of peace and goodwill towards others.

Chrysocolla Mystery surrounds the chrysocolla, reputed to have been used as a healing stone by the priesthood of ancient civilizations. No record exists of the way in which the stone was used or of the reason for its reputation as one of the most efficacious of the healing stones.

Chrysoprase Perhaps the ideal stone for those about to take tests and examinations, the chrysoprase is said to improve the memory and reduce nervousness and impatience. It is also reputed to increase the presence of mind of the wearer and make him more able to act with sound judgement in an emergency.

Citrine It is claimed that the citrine is able to help those who feel that they have misplaced their path in life and to give them a new sense of direction. It is also said to be of benefit to those suffering from poor circulation and to help to control the emotions.

Coral Coral is said to promote general physical and mental well-being and to be of particular assistance to those suffering from anaemia. In many parts of the world it is believed that the coral can be used to ward off evil thoughts sent by ill-wishers — and indeed there are areas where it is still used in this way.

Diamond The diamond is supposed to be of greatest benefit when used in conjunction with other gems, as it is claimed that it enhances the properties which other gems contain.

Emerald The emerald is said to improve both the memory and the intellect, as well as to be of assistance in overcoming feelings of depression and insomnia.

Garnet The garnet is most often used as a general tonic for the whole system — physical, mental and emotional. It is

Healing

	particularly recommended for those who need to improve their self-respect and self-confidence and to increase their courage.
Haematite	The haematite is another stone which is said to increase courage and it is also claimed that it strengthens the heart and is good for reducing a rapid pulse.
Hawk's eye	Not the most common of stones, all that can be discovered about the claims made for the hawk's eye is that it is believed to be helpful in overcoming eye diseases. So little information exists, however, that one cannot be sure the reputation does not arise merely because of the stone's name.
Jade	Jade is said to be of help in relieving kidney complaints. Yellow jade is believed to aid a poor digestion. When worn as a piece of jewellery, jade is thought to provide protection from one's enemies. In ancient China and ancient Egypt it was widely used as a talisman to attract good fortune, friendship and loyalty.
Jasper	Green jasper is reputed to be able to improve the sense of smell and also to overcome depression, bringing stillness to a troubled mind. Red jasper is known to contain iron oxide which is used medically to control excessive bleeding and for this reason, it is claimed, it can be useful in overcoming disorders of the blood and even in reducing a tendency to haemorrhage.
Jet	Jet is actually a fossilized plant and not a stone at all. None the less it is used in healing to control and ease migraine and pain behind the eyes.
Lapis lazuli	The lapis lazuli was called by the ancient Egyptians 'The Stone of Heaven' and it is thought by many to be the stone upon which were carved the laws given to Moses. It is said that it has the power to prevent fits and epilepsy and to improve the eyesight.
Lodestone	Also known as the 'Hercules Stone', the lodestone is reputed to be of assistance in healing such ailments as rheumatism, gout, neuralgia, cramp and poor circulation — particularly in the legs and feet.
Magnetite	This stone is believed to help in the treatment of rheumatism, liver and eye diseases.

Malachite	The malachite contains copper and is claimed to be helpful in the treatment of rheumatism and also in regularizing menstruation.
Moonstone	The moonstone is claimed to promote long life and happiness and is said to attract friendship and loyalty towards the wearer. In healing terms, it is often used to reduce excess fluid in the body and to reduce the swelling caused by this fluid.
Obsidian	Obsidian is actually volcanic lava and is said to be good for improving the eyesight. Mystics also believe that it is beneficial in improving the higher vision of spiritual awareness.
Onyx	The onyx is reputed to improve concentration and devotion. Perhaps it is for this reason that it is frequently to be found in rosaries.
Opal	To some the opal has the reputation of being an unlucky stone. This could actually be because it is supposed to induce daydreaming in the wearer — and daydreaming of course can be detrimental if it causes the loss of a sense of reality. Daydreaming, however, can also equal inspiration — and the dreamer may spend his 'lost time' formulating plans and ideas which will have rewarding results, or visualizing a piece of creative work which will later bring beauty to others as well as himself.
Pearl	The pearl is said to be helpful in clearing all forms of catarrh, bronchitis and chest and lung infections. It is still traditionally worn by divers to protect them from the evils of the sea — especially sharks.
Peridot	The peridot, as well as being recommended as a cure for insomnia, is said to aid the digestion and be useful in reducing fever.
Pyrite	Pyrite is claimed to increase the oxygen supply in the blood, to strengthen the circulatory system in general, and to be useful in clearing congested air passages.
Quartz	The quartz attracts the powers of light and energy and is said to be an excellent general healer. Its greatest attribute is known to be its use as an aid to opening the psychic centres, enabling one to meditate at a deeper

	level and to free one's mind from the mundane and from trivia. It releases the higher consciousness of the individual and assists in the development of many mystical and spiritual gifts.
Red Coral	Red coral is reputed to be of benefit in the treatment of liver disorders, constipation, eczema and other skin troubles. It is said to cleanse the entire system and strengthen mental faculties.
Rhodochrosite	This gemstone is being used more and more frequently to integrate the physical, mental and emotional energies of the user.
Rose quartz	This lovely stone is claimed to be one of the best stones to use in the treatment of migraine and headaches of all types. It is also said to stimulate the imagination and the intellect.
Ruby	As well as aiding intuitive thinking, the ruby is believed to increase levels of energy. It is often used to alleviate disorders of the blood, such as anaemia, poor circulation and menstrual problems.
Rutilated quartz	The rutilated quartz is said to be of particular benefit to those who suffer from respiratory complaints of any form — such as asthma, bronchitis, etc. It is often used in conjunction with the beryl.
Sapphire	The sapphire is reputed to have a great number of healing properties. When used on the physical body it is said to help to overcome such problems as backache, skin eruptions, the condition of the hair and nails and also to prevent bleeding. There have also been claims that it can be beneficial in the treatment of various forms of cancer. On an emotional level, the sapphire is said to promote an intensity of loving feeling and to give the wearer the qualities of religious devotion, purity of mind and serenity.
Serpentina	It is believed that the serpentina increases the wisdom and the self-control of the one who wears it. Tradition has it that this stone should always be worn attached to a cord rather than to a metal chain.
Smokey quartz	For centuries the smokey quartz has been used as a talisman or a good luck charm. It is thought that its

	beneficial qualities lie rather more in this direction than in healing itself.
Sodalite	Blue sodalite is reputed to assist in the lowering of blood-pressure and also to have a cooling effect upon those suffering from a fever or an over-high temperature.
Spinel ruby	The spinel is said to increase the character and the strength of purpose of one who wears it and also to attract help and guidance when it is needed.
Staurolite	Also known as the 'Fairy Stone', the staurolite is claimed to be of benefit to those who feel generally below par but cannot identify the reason for this. It is also thought by many to be a stone bringing good fortune, but this could be because of the markings of the cross which can be seen on it.
Tiger's eye	It is claimed that the tiger's eye will counteract feelings of hypochondria and the onset of psychosomatic illness, and will also give a feeling of self-confidence and belief in himself to the wearer, thereby inducing increased self-knowledge. This stone is also believed by some to be of assistance in healing eye diseases, but there is little support to be found for this claim and it could be that the reputation arose merely because of the name of the stone itself.
Topaz	It is said that the topaz can relieve high blood pressure and reduce varicose veins. It is also believed to prevent insomnia and to encourage sound dreamless sleep. There are also some who think that the topaz can prevent baldness — but very little substantiation has been found for this claim.
Tourmaline	The tourmaline should be worn against the skin for maximum effect. Its particular properties are reputed to be the relief of nervousness in the wearer and the encouragement of self-assurance. It is, in fact, also known as the 'confidence stone'.
Turquoise	Some American Indians still consider the turquoise to be a sacred stone. They believe that it absorbs harmful vibrations and that it is a protective stone. A turquoise — which should always have been given and never bought for oneself — was at one time frequently given to those

who were about to undergo surgery as a form of protection during the operation.

Zircon It is believed that the zircon is helpful in increasing the appetite and in overcoming problems of the liver.

3.

The Chakras

The chakras are the centres of energy in the body of every human being. There are seven major chakras in the human body and these are often depicted in diagrams as wheels or discs. Indeed, the word chakra means 'wheel' in Sanskrit.

The seven chakras of the human body are:

1. **The Crown Chakra** — situated at the top of the head;
2. **The Third Eye/Brow Chakra** — situated at the central point between the eyebrows;
3. **The Throat Chakra;**
4. **The Heart Chakra** — situated in the centre of the chest area;
5. **The Solar Plexus Chakra** — situated at the level of the navel;
6. **The Spleen/Sacral Chakra;**
7. **The Root/Base Chakra.**

Each chakra is responsible for controlling and relaying information to a different area and, although as will be seen they function from a central force and have a degree of interdependency, none the less each is responsible for the actions and reactions of a different part of the mind, the emotions or the body. Naturally one is using all the chakras all the time, but which one predominates at any given moment will depend upon the individual circumstances applying at the time. For example, you may be in a situation where you need to use the mind and the intellect above all else — this does not mean that the other chakras will not be functioning, merely that it will be those which control the mental and logical aspects that will be the ones most in evidence.

The chakras were earlier referred to as 'wheels', and what an appropriate image that is. Just like the cogs of a wheel in any other elaborate piece of machinery, the various chakras interrelate and the efficiency of each is dependent upon the successful functioning of the others. Because energy is transferred from one chakra to another in an ever-constant interchange, if there should happen to be a blockage in the area of a single chakra it can lead to inadequate functioning of the others, which in turn will cause an imbalance in the bodily functions.

The Chakras

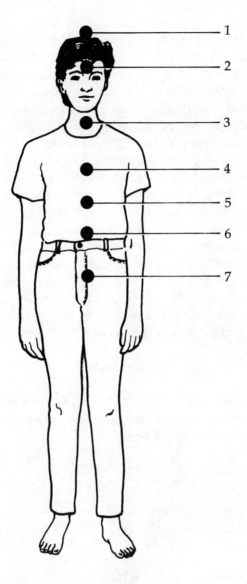

Figure 5: The seven chakra points

Your life force enters through the crown chakra and it activates each chakra in turn. Suppose there were to be a blockage in the heart chakra; insufficient energy would reach the solar plexus, spleen or base chakra, thereby causing an imbalance in the functioning of these three.

Blockages are caused by negative thought which may not, of course, be on a conscious level but purely subconscious. Those blockages can be cleared by deliberate application of positive thought. Quartz crystal, which is capable of amplifying energy, intensifies the level of positive thought.

The object of this chapter is to help you to recognize those blockages in yourself and to teach you how to set about removing them. Even if you are not aware of blockages, the following exercises may enable you to become more sensitive to your mental, physical and emotional functions and more aware of how these functions are controlled by the chakras. You are never going to inflict harm upon yourself by following the techniques set out here and observing the results and it may be that you will be able to improve both your appreciation of the quality of life and your ability to be of assistance to others.

The crystal you should use for cleansing and activating the chakras should be a small one — not more than two or three inches in length. The chakras are highly sensitive, and a large crystal would amplify energy to such an extent that it would be much too strong a force for you to cope with.

Now, if you are ready to begin, let us start with the crown chakra and learn how to activate it. At the end of each of the seven sets of information, you will find a chart on which you will see the normal reactions to be expected if this process is working successfully. You may find it helpful to make note of your own reactions on these charts. Please do not be disappointed if you do not experience all of them each time you practise. Take your time and eventually you will feel all that you are supposed to. In addition, you may experience sensations which are not itemized. There is nothing wrong in this but it is as well to make a note of any feeling which may occur.

Crown Chakra

The crown chakra is linked to the pituitary gland and if this chakra is properly balanced there will be harmony throughout the entire glandular system. This in turn promotes general physical health.

To clear and activate

Figure 6: Activating the crown chakra

Hold your crystal in your right hand a few inches above your head with the point downwards. Keeping your eyes closed, rotate the crystal gently in small clockwise circles and, at the same time, visualize strong white light emanating from the crystal and entering the top of your head. Continue for some moments, still rotating the crystal. Then relax, lower your arm and sit quietly for a few minutes allowing the visualization to continue.

Crown chakra experiences			
Discomfort			
Heat at the point of entry of the light			
Tingling sensation			
Flashing coloured light			
Patterns and designs — describe			
Any other sensations — describe			

You will be unlikely to experience the patterns and designs mentioned above until you have been practising this technique for some time but, as you may be opening your channel of clairvoyance by activating this chakra, it is important to make notes of any images, patterns, etc., even if you are not able to understand them at the time you first see them.

Third Eye (or Brow) Chakra

The third eye chakra is linked to the pineal gland. When this chakra is functioning correctly, you will find that you are less prone to fears and phobias. In fact, if you are a person who experiences a great deal of fear in your life, that is a strong indication that you do have a blockage in the area of the third eye chakra. The third eye, of course, is more often recognized in the context of psychic or spiritual ability and you will indeed find that, as your brow chakra becomes clearer you will become more psychically aware than you had been before. This is an aspect of human capability which is insufficiently used today and we can only help ourselves and others by allowing it to develop.

It is said that the Atlanteans always wore a gemstone over the point of the third eye chakra to inspire perception and psychic ability. Of course there are some Eastern countries where a jewel is worn in the centre of the forehead today but it is possible that the true reason for this has been lost and that it has become merely a matter of tradition.

To clear and activate

Hold your crystal in your right hand about three or four inches away from the point of the third eye chakra. The point of the crystal should be towards your

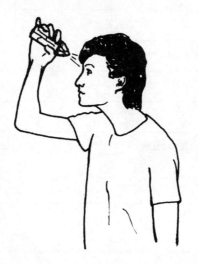

Figure 7: Activating the third eye chakra

brow and the angle of the crystal slightly downwards. Close your eyes, rotate the crystal gently in a clockwise direction for several minutes, and then relax and allow the sensations to continue.

Third eye or brow chakra experiences			
Slight 'stabbing' sensation			
Heat			
Circle of light — may be white or coloured			
Any other sensations — describe			

The clearing and activating of the first and second chakras (the crown chakra and the third eye chakra) are often done in conjunction with one another as the two chakras have very strong links, particularly in relation to increased awareness and to psychic and clairvoyant abilities. Harmony between these areas is therefore of the utmost importance. Indeed, as these chakras become clearer, you are certain to find that you are in receipt of images and information which you may well wish to record. These images may not appear to make much sense in the beginning as it is possible that they will be symbolic in nature rather than literal, but notes kept over a prolonged period of time should enable you eventually to understand. In this way you will reach a greater understanding of yourself and will be more able to offer help and guidance to others. Once again, it is important that you do not expect too much to happen at once. Any increase in ability and awareness is bound to take time and everyone will proceed at a different pace — there is no correct time limit to be set upon development of this nature.

Throat Chakra

The throat chakra is linked to the thyroid gland. As is quite widely known, the thyroid is linked in turn to the body's metabolism. Stress, anxiety and fear may easily interrupt the smooth functioning of this chakra and it may take some time to clear as, in this twentieth-century world, stress and its effects are prevalent.

Figure 8: Activating the throat chakra

To clear and activate

Hold your crystal in your right hand, no more than six inches from the throat chakra. Make certain that the crystal is in a horizontal position, parallel to the floor. Close your eyes and visualize a jet of pure white light emanating from the crystal and entering your throat at the position of the chakra.

Throat chakra experiences			
Heat			
Desire to cough			
Feeling that the body's 'engine is racing'			
Increased energy			
Any other sensations — describe			

Heart Chakra

The heart chakra, which is linked to the thymus gland, is not located in the area of the heart itself but in the area at the centre of the chest. There are not many people who have a fully activated heart chakra as this is the area from which unconditional love begins and very few are able fo feel this true and

Figure 9: Activating the heart chakra

selfless love. That does not mean, however, that it is not something we would all strive for — and how much better a place the world would be if we were all able to achieve it.

It is also said that, if the heart chakra is activated, the body will increase its resistance to infection and its ability to overcome the effects of allergies.

To clear and activate

Hold the crystal in your right hand with the point towards the heart chakra and not more than three or four inches from your body. The crystal should once again be parallel to the ground. Close your eyes and visualize both a strong white light coming from the crystal and entering your body at the level of your heart chakra and, at the same time, a pure white light coming from the area of your heart chakra itself and linking with the light emanating from the crystal.

Heart chakra experiences			
Tingling sensation			
Warmth			
Sometimes tears (not a bad sign)			
Any other sensations — describe			

Solar Plexus Chakra

The solar plexus chakra is strongly linked to the individual ego. As it is human nature to pursue and follow the desires of the ego, it often takes repeated attempts to cleanse and activate this chakra effectively. It is easily and frequently affected by any friction or trauma, either from the inner self or from an outside source. When you consider just how much friction we are subject to in our lives, you will understand the importance of the solar plexus chakra and of the necessity to work on it often and well. It is an area which is the subject of so much conflict within us that even the majority of those who try to increase their own self-awareness and unselfishness in order that they may grow spiritually have to strive hard to overcome the natural selfishness which is inherent in man.

The solar plexus chakra is also linked to the pancreas, responsible for the production of insulin and glucagon. Should the pancreas fail to function effectively, sudden changes in weight, or even diabetes, may occur.

To clear and activate

Hold the crystal in your right hand with the point towards the solar plexus chakra. Make sure that the crystal is completely horizontal and parallel with the ground. Close your eyes and visualize a strong white light entering your

Figure 10: Activating the solar plexus chakra

Solar plexus chakra experiences			
Throbbing/pulsating			
Feeling of heat			
Release of tension from spine			
Any other sensations — describe			

body at the level of the chakra itself. Do not be disappointed if you feel little or no response on the first few occasions you try this exercise. Working on this particular chakra requires perseverance on your part — and even then results may come only slowly over quite a long period.

Spleen Chakra

The spleen chakra is linked to the body's production of adrenalin which concerns stress and how the individual copes with it. A certain amount of stress in your life is necessary and even stimulating. It is when that stress becomes excessive that it can cause innumerable problems, ranging from headaches through to the truly debilitating diseases — physical as well as mental. If the spleen chakra is not properly balanced, the equilibrium of the entire system of the body may be affected. Because we all have to deal with stress in our everyday lives, it is vitally important to maintain the correct functioning of this particular chakra.

The spleen chakra is also linked with healing and with feelings of love for mankind. If, therefore, it is your desire to become a healer and to help others, it is essential that you work regularly on this area.

To clear and activate

Hold the crystal in your right hand pointing slightly downwards towards the area of the spleen chakra. Close your eyes and rotate the crystal gently in a clockwise direction. At the same time, visualize a pure white light entering your body at the position of the chakra and then spreading outwards, like the ripples on a pond, until it fills the whole surrounding area.

Figure 11: Activating the spleen chakra

Spleen chakra experiences			
Little or no reaction			
Warmth			
Tingling in area of chakra			
Relaxation of spine			
Serenity of mind			
Warmth spreading thoughout body			
Any other sensations — describe			

Base Chakra

The base chakra is situated just above the reproductive organs and relates to the lower instincts and sexual activity of the individual. If the equilibrium of the base chakra is not maintained, it can lead to excessive sexual appetites, frigidity, infertility and even, in some cases, mental instability.

To clear and activate

Hold the crystal in your right hand and point it downwards towards the position of the base chakra. Close your eyes and visualize the bright white light entering the body at a single point. Hold the crystal steady when working on this chakra; do not rotate it.

Base chakra experiences			
Heat at the point of focus			
Spreading of heat throughout body			
Muscle relaxation in lower body			
Release of tension in body			
Any other sensations — describe			

Figure 12: Activating the base chakra

These seven exercises should be performed on a regular basis until such time as you are able to feel a real response in the area of each chakra. It is also important to try to become aware of how the stimulation of one chakra affects all the others and of the changes in your attitudes to life and your general increase in awareness. Because these things are so easily forgotten, it is a good idea to keep some sort of diary or journal in which you can make notes of the responses you feel.

Just as anything in life which is worth having takes time and effort to achieve, do not expect miraculous results within a very short period. Depending upon your nature and the point from which you start, you will find some chakras easier to activate than others. It is important, however, to continue to work upon all of them as it is their interreaction and their effect upon your entire life which are important.

The opening and closing of the chakras is also a significant part of psychic development, which we will deal with further in a later chapter.

Once you start to clear the chakras of negative energies, it becomes necessary for you to use a form of protection regularly so that these newly-activated chakras do not too easily absorb the negative influences which abound in the ordinary world. Each time you have practised these activating techniques, take a few moments at the end of the session to relax and to visualize yourself surrounded in that same pure white light which has been emanating from the crystal and entering the chakra points. Start from the ground and work upwards, allowing this pure white light to completely cloak you in its protection and ensuring that this protective shield seals itself above your head. You can then continue with your everyday life without the vulnerability which you might otherwise experience.

4.

Psychic Development

Before you begin to read this chapter and practise the techniques set out here, it is essential to understand that there is a difference between development as a psychic and development as a clairvoyant or a medium. The task of the clairvoyant or the medium is to give proof of survival to those who seek it. This will be dealt with in more detail in a later chapter. What we are dealing with here is those who wish to learn to use their inherent psychic abilities.

The psychic, once he has developed, may then go on to use any of a number of 'tools' to help him in his work. One may choose to use Tarot cards, while another may prefer the crystal ball. Later in this book you will be taught how to use crystals and semiprecious stones in a psychic reading. It does not matter at all which method the psychic decides to use; those 'tools' will be for him a focal point — something upon which to concentrate while allowing his own intuition to guide him. That is not to say that Tarot cards, semiprecious stones, etc., do not have meanings of their own, but those basic meanings will always be open to more than one interpretation. If the psychic has developed sufficiently, it will be his own intuition which tells him which interpretation is the right one in each case. In addition, when you reach this stage in your development, you will find that a point is reached in the course of each reading when you are no longer using your physical 'tools' but you are giving guidance and information which you have received on a psychic level.

A good psychic should be able to tell the questioner something about his past, his present and his future. It is never the task of the psychic to give the questioner instructions as to how he should run his life as this would be to take away his freedom of decision. It is for the psychic to point out the possible paths and, if necessary, any areas where the greatest pitfalls may arise. Many of those who consult a psychic do so because they are at a low ebb, or because they feel that they have lost their hope. The best psychic must consider himself a counsellor too and should do his utmost, while not lying or masking the truth, to send his questioner away with hope in his heart and with positive feelings about the future. If you aspire to help others by means of your own psychic ability, do not forget that you have a great burden of responsibility; you are dealing with the hopes, fears, emotions and desires of another human being and one who may be experiencing feelings of negativity when first he comes to consult you. Remember too that there is no such thing as predestination and that the questioner must make his own decisions and choose his own path. It is your job to show what those paths may be and the decisions between which he has to choose, but *never* to make his final decisions for him.

Everyone has the inherent ability to be psychic or intuitive. Not everyone realizes this — and the vast majority of people have no interest at all in this area of personal development. Of those who are aware enough to realize that such development is possible, some will try to find methods of training whatever talent they already have. It is rather like the ability to sing. Everyone can go 'la-la-la' more or less in tune. Everyone can be trained to sing rather better than they do, although only some will become the Pavarottis of the world. What matters is the recognition of the basic talent and the desire to improve.

Should you wish to develop psychically, you can do so either alone or in a group. If you choose to develop within a group there are a few important points to remember:

1. The group should have a leader who is already an experienced psychic or medium in order that each individual can be protected from any negative influence which may have been brought to the group — however inadvertently — by one of the members.
2. The group should not be too large. The more people involved, the more likely it is that one of the members will be going through a negative period in his own life, and this can adversely affect the other members of the group.
3. You must feel that you are compatible with the other members of the group and that you are meeting for a common purpose. There will always be those people who would like to feel that they are 'special' and whose motives for joining the group are entirely selfish, and this could hamper the progress of the other members.
4. You must feel an affinity with the group leader. If this person is to help you in your own development, it must be someone whom you like and trust as you will be putting yourself in quite a vulnerable position and must be able to rely upon his sense of purpose and his strength.

Whatever method of development you choose, whether it is to increase your own awareness by yourself or to join with others within a group, crystals can make the progress easier for you. Whatever inherent psychic abilities you possess, crystals can only serve to enhance them. It is not being suggested that the use of crystals will make other forms of development unnecessary, merely that it may serve to intensify and accelerate the process.

Before you begin to develop your own psychic ability, there are three things to consider:

1. It is necessary to be aware of and to accept your own inherent psychic awareness, however unchannelled it may be at the present time.
2. The desire to develop must be in *you*. Although you will receive help and guidance, there is no 'magic wand' to be waved and you will have to work hard to achieve any real level of ability.
3. What is your reason for wanting to develop? Hopefully it is because you wish to help other people and to become more spiritually aware for unselfish reasons. Having said that, you will actually gain a great deal of personal satisfaction from being able to be of assistance to others — but this satisfaction itself should not be the prime motive for development. There are those who wish to increase their psychic abilities for entirely the wrong reasons — for personal power over others, for example. Be assured that the spiritual laws and the laws of karma will always see to it that such people will have to pay the penalty for the misuse of their psychic talents. This is not to suggest that it is wrong to earn a living by using your psychic abilities to help and advise people, any more than it is wrong for any other type of counsellor to earn his living by giving help and advice in the way in which he has been trained.

One word of caution: it is not advisable to begin your own psychic development at a time in your life when you are feeling depressed and unwell. It will be far harder for you to accomplish anything of significance and you may well drain the energies you need to improve your own condition. If you are feeling depressed or physically unwell, you would do better to seek help in improving that situation before starting your own development.

You will find in this chapter various techniques for development, either alone or in group situations. It will be necessary for you to try any of them which may appeal to you — but please take your time. There is no point in trying method one on the first day, method two on the second, and so on. Each exercise should be practised for a period of two to three weeks, during which time you should keep note of how easy or difficult you found it to do, whether you felt that it was a method which appealed to your personality, and what results, if any, were achieved. Only after you have done this with each exercise in turn, will you be in a position to compare them and decide which one (or it may be more than one) you would like to concentrate on to help you in your own development.

Although there is no doubt that, at some time in your life, you will have had things happen which you had 'foreseen' or you will have 'known' something about another person without that person actually telling you, many of you will be wondering whether it was 'just imagination' or 'coincidence'. It is not a bad thing to wonder — it shows that you are not so conceited or full of yourself that you automatically assume that you have this special psychic ability. And it is only those who are sufficiently humble by nature who are suitable to use their talents, once developed, to help other people. Once you have accepted that you have this inherent ability and that there is, in fact, no such thing as 'coincidence', you may go on to wonder whether you are foolish in trying to develop further and whether you will not be able to achieve anything worth having. It is important to remember one point here: whether you think of yourself as being helped by God, guides or Spirit (or any other word you may care to use) — why should that special 'something' have brought you to the edge of development only to allow you to fail or to make a fool of yourself? So, if you feel unable yet to trust yourself and your own abilities, put your trust in that power much greater than you, the one which is assisting you to develop sufficiently to be of help to your fellow man.

Whichever of the techniques set out later in this chapter you decide to begin with, there are a few prerequisites to be met first:

1. Try to set aside a regular time for your development. It is not important whether it is daily, twice a week or even weekly — it is regularity which matters.
2. Always use a crystal which is your own and which has been properly cleansed and charged. The crystal you use to help you in your psychic development should be kept for that purpose alone and not used for anything else.
3. Before you begin any exercise, take the time to ask for guidance and protection from the world of spirit.
4. Never fail to spend a few moments making sure that you are surrounded by that pure white light of protection. To open yourself up to whatever forces may be around without first ensuring that you are safely protected is as foolish as going out for the day and leaving your front door open. You could be fortunate and a friend could enter your house to make sure that all is well — but you could also be unlucky and return to find that your house has been vandalized. So it is absolutely essential that, before you open up any of your psychic centres, you see to it that you are well protected.

Development Exercises For One Person

Exercise One

Once you have had some experience of opening and activating the chakras (as explained in the previous chapter), relax, holding your chosen crystal in both hands, the point uppermost. Visualize the chakras opening one at a time, starting with the base chakra and working upwards. Once you have reached and opened the crown chakra, remain in the same position for ten or fifteen minutes and be aware of any sensations you may experience, whether they be visual, audible or symbolic. Before opening your eyes and making note of your experiences, be sure that you visualize all the chakras closing down one at a time, starting with the crown and ending with the base chakra.

Exercise Two

For this exercise you can use either a quartz crystal or an amethyst. Close your eyes and relax, ensuring that your breathing is smooth and even. Imagine the chakras opening in the usual way, starting from the base and working upwards. Then hold your chosen crystal gently against the point of the third eye. Hold it there for a few moments before lowering your hand and relaxing — still holding the crystal in your hand. Remain in this position for ten or fifteen minutes and notice anything you may experience. Once again, before you open your eyes, it is vitally important that you close down all the chakras, just as before, so that you do not leave yourself in a vulnerable state once the exercise is completed.

Exercise Three

Lie on a bed or on the floor with four or five *small* suitably charged crystals in a semi-circle around your head, the points towards you (see Figure 13). Place a charged amethyst in the position of your third eye, close your eyes and relax. Remain in this position for ten or fifteen minutes. (You may find that on the first few occasions you try this particular exercise all that you experience is flashing coloured lights but this is quite normal and does not mean that you are doing anything wrong.) Remove the amethyst from your forehead, open your eyes and sit up gently.

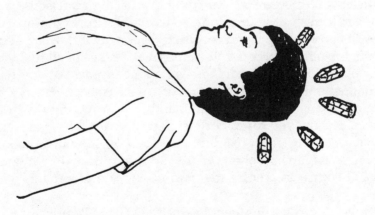

Figure 13

Exercise Four

Holding a large, charged crystal in both hands and resting your hands in your lap, sit quietly with your eyes closed. Visualize the opening of the chakras and then remain in a relaxed and meditative state for some minutes before opening your eyes, lifting the crystal to a comfortable position and gazing into the heart of the crystal. Really focus your attention on the inner depths of the crystal and take notice of what images appear. Sometimes they will be vague and sometimes the 'picture' will be extremely clear-cut and easy to recognize. As soon as you find your attention wandering, lower the crystal to your lap again, close your eyes and visualize the chakras closing down, starting with the crown and ending with the base chakra.

Exercise Five

Sit on the floor, surrounded by a circle of suitably charged crystals. Close your eyes and visualize the opening of the chakras one by one. Take yourself on an imaginary 'walk' in your mind. The walk can take place somewhere with which you are familiar or it can be a spot which is created solely in your imagination. Make sure, however, that it is the type of walk you would enjoy — whether it is by the sea, in the mountains, or in a country lane. Although you will start by following a path in your mind, you will find that after a few moments seemingly unrelated images will appear. Do not try to prevent this but merely become an observer, seeing what happens but not forcing the images in any way at all. Allow this to continue for a comfortable period of time before relaxing, closing down the chakras and opening your eyes.

Whichever exercise you choose to practise, it is just as important that you offer up thanks for the guidance and protection you have received as it was you who asked for it in the first place, so never forget to do this.

Try to make notes of all that you have seen and experienced during the development exercise as soon as possible after the exercise is completed and while it is still fresh in your mind. You will be surprised at just how quickly images will become vague and confused, especially if they are symbolic rather than factual. Do not worry if you cannot make sense of all that you have seen — it may be that the true meanings will not be revealed to you for some time. In addition, should you not be aware of any sensations or any experiences at all, do not feel that you have failed in any way. Anything worth having takes time to achieve and you cannot expect miraculous results every time. But even if you have nothing of any significance to write in your notebook, you will have spent some time becoming more and more attuned with your higher self and this, of course, is an essential part of your development.

Development Exercises For Two People

In order for these exercises to work successfully, it is essential that the two people feel comfortable with each other and are working towards the same end, and also that the crystals they use should be compatible.

Sometimes two people who choose to work together for their psychic development will use two halves of the same crystal in the exercises. If you choose to do this you must be sure that each half of the crystal is recleansed and recharged after cutting and before use.

Exercise One

Sit facing one another, each holding your own crystal. With your eyes closed, open your chakras one at a time. When you have completed this process, each person should hold his crystal gently against his forehead in the position of the third eye. After a few moments each person should relax and hold his crystal gently in his hand, with that hand resting in his lap. Remain like this for ten or fifteen minutes, before closing the chakras and opening the eyes. You will be surprised, when comparing notes afterwards how often the two of you will be aware of identical or complimentary sensations and images.

Exercise Two

Let one person be seated in a straight-backed chair and the other stand behind him. Both should close their eyes and the person who is standing should rotate one small crystal slowly in a clockwise direction above the head of the person seated. At the same time he should hold another small crystal gently against the position of his own third eye. Although only sensations of light or of colour may be experienced on the first few occasions that this exercise is practised, eventually the person who is standing should receive information or images which have a particular significance to the one who is seated. Be sure, of course, that the crown and third eye chakras of each person are closed after the exercise and before opening the eyes.

Exercise Three

To see if you have a telepathic link between you, each person should be seated, with his eyes closed and facing his partner. Holding his crystal in both hands, each should concentrate on opening his chakras, working as usual from base to crown; you should then visualize a strong white light between the third eye of each of you, linking those chakras. One person should concentrate on a particular image (of which the other person has no advance knowledge) for some moments, while the second person should allow his mind to be receptive in order to see whether he is able to receive that same image from his partner. Make sure that you relax and close down the chakras before opening your eyes and discussing what you have experienced. Do not be disconcerted if you are not able to transmit or receive specific images when you first try this exercise. Even those who have later proved to be extremely telepathic have had to practise to reach this state and did not achieve it on the first attempt.

Development Exercises For Group

If there is an experienced member of the group, it is often helpful if he gives directions to the other members (for example, about opening and closing of the chakras) so that there is some sort of synchronization about the exercise.

Exercise One

Sit in a circle, each member of the group holding his own suitably charged crystal. Close your eyes and let each member of the group spend some time in silent meditation, opening his chakras in the way already explained — always starting with the base chakra and ending with the crown chakra. Once this has been done, each member of the group should visualize a strong white light travelling in a clockwise direction around the circle several times. After this, remain quietly in the circle for about ten or fifteen minutes while each person becomes aware of any sensations or images which he may experience. Before opening your eyes at the end of the exercise, be sure that each person closes down the chakras again, working from the crown to base.

Exercise Two

Sit in a circle around a group of large crystals — using either quartz or amethyst (but not a mixture of the two). Let each person close his eyes and concentrate upon opening his own chakras — starting, as always, from the base chakra and working on each until you reach the crown; relax and concentrate upon the rhythm of your breathing and upon the feeling of harmony among the members of the group. Visualize a strong ray of light (white if using quartz crystals or purple if using amethysts) coming from the group of crystals in the centre of the circle and entering the brow chakra at the position of the third eye. Allow yourself to become aware of the sensations you experience. In a group development of this particular type, you may well find that you experience feelings and emotions rather than actual images. After about ten minutes, each person should relax and take time to close down his chakras before opening his eyes.

Exercise Three

Place one large crystal in the centre of the room and sit in a circle around it. Let each member of the circle hold his own personal development crystal in both hands. After closing the eyes and opening the chakras in the usual way, each member of the group should raise his personal crystal to the level of the third eye and gently touch the appropriate point on his forehead. Hold this position for a few moments before lowering the crystal again and allowing the mind to concentrate on whatever sensations may arise. Remain in this position for about ten or fifteen minutes before closing down the chakras and opening the eyes.

It is just as important when working with others in this way to make notes of what you experience, both individually and as a group. One of the advantages of working together is that there is an opportunity to discuss what each member has experienced and to try to understand the meanings of those images which are symbolic in nature. If, as is desirable, there is an experienced member of the group, he may be able to give some guidance as to the possible interpretation of the images received.

If development is to take place within a group, it is usual for one member of that group — not necessarily the one who is most experienced — to say the opening prayer, asking for help and guidance and also to offer thanks at the end for the guidance and protection given.

You now have several exercises to try in order to assist your own psychic development. You will feel a real affinity with some and not with others — and that is as it should be. We are all different. Our crystals, although each possesses the ability to increase the energy we receive, are all different and each has its own personality. That does not mean that you would not be well advised to try each of the exercises for yourself and see which of them you feel most comfortable with.

A few reminders to help you when you practise the various exercises for psychic development:

- Always begin by asking for help and guidance in your own development;
- Never start an exercise without first making sure that the white light of psychic protection surrounds you;
- Open and activate the chakras from base to crown and, at the end of the exercise, close them in the reverse order;
- Be sure that the crystals you use have been properly charged and cleanse and charge them frequently;
- If working with a group (or with one other person), be sure that all have the same motives and are working towards the same unselfish end;
- Never forget to end the exercise by offering thanks for any help, protection and guidance you have been given;
- Practise each exercise for at least two weeks before proceeding to the next. Do this even if you feel that a particular exercise is not suitable for you. You may be pleasantly surprised with the results you achieve after several days of practice;
- Always keep notes of what you have seen, felt and experienced even if you are unable to understand the meaning of the images. True meanings may not be shown to you for some considerable time;
- Above all, do not be despondent if you feel that you are not progressing as quickly as you would like or if you are not receiving clear and meaningful images and messages. Remember that anything worth having will take time to achieve and every step of the way is significant, even if that significance is not easily recognized.
- Enjoy the exercises. You have chosen to increase your awareness and your

sensitivity and, provided this is being done for the right motives, you will receive the necessary help and guidance. So practise with joy in your heart, secure in the knowledge that you are doing your best to progress and that the power and energy inherent in the crystals will help to accelerate your development.

5.

Progressing Towards Mediumship

The title of this chapter has been very deliberately chosen. What is set out here is intended to help you to make progress should you wish to develop mediumistic abilities — but it is only meant to act as a stepping-stone on your journey. Full development should only be undertaken under the guidance of an experienced and competent medium who will be able to help you on a more personal level. That development may take place on a one-to-one basis or, as is more usual, in a development circle under the guidance of a medium. Whichever method you may eventually choose, if you spend some time practising the technique set out in this chapter, you will already have begun to develop your own spiritual awareness and hopefully will find the path to mediumship easier and more rapid than you would have done had you been a complete novice.

It is important at this point to stop and consider what is meant by mediumship and to distinguish between this form of awareness and psychic development as outlined in the previous chapter. Mediumship involves contact with spirits who have departed this earthly life and have passed over into the life beyond. A professional medium aims to give evidence of survival of these spirits, as well as to give help, comfort and peace of mind to those whom they contact.

There are, in fact, many forms of awareness. Not every medium will be clairvoyant — that is, one who actually *sees* what he is describing; not every medium will be clairaudient, hearing voices and words — some excellent mediums have only been able to express the sensation they experience as 'knowing' what is there and being able to describe it even though no visual or audible image may appear. None of these methods is better than any other, and it is only by experiment that you will discover which type of medium you are meant to be. Some mediums, and you may well be one of these, have more than one form of awareness and can 'see' as well as 'hear' those things of which they are giving evidence.

Whichever type of mediumistic abilities are to be yours, your journey towards the attainment of those abilities can be accelerated and assisted by the use of quartz crystals. A quartz crystal, as we have already seen, amplifies energy and, as it is thought energy which is used in mediumship, the additional power of crystals can be a great help.

For those who wish to develop as a medium, it is of vital importance that the correct crystal is selected. Ideally you should have several from which to choose. Hold each crystal in turn and take note of your reactions. If you feel a tingling sensation in the area of your forehead or the top of your head, then you have found a crystal which would be ideally suitable for you. (It is as well to wipe your hands before and after handling each crystal as otherwise what you feel may be an accumulation of energy built up after holding a number of different crystals and this is not what we are looking for.)

When you have chosen the crystal you intend to use to aid you in your development, it is obviously essential to cleanse it and charge it as you have already learned. Once this has been done, the crystal should be kept completely separate from any other crystals you may have and should on no account be handled by others.

If you use a quartz crystal in your development you will find that your spiritual awareness increases quite rapidly. It is important to establish a routine and to practise at regular intervals. Try not to be unrealistic. There is no point in setting yourself targets you cannot hope to keep. It is far better to say that you will practise once a week and to keep to this routine than to promise yourself that you will set aside a time for practice every single day, only to find that circumstances do not permit this. What follows below is the basic technique for the early stages of development towards mediumship and it is this technique you should try. Make note on each occasion of any impressions you may receive — and do not be too disappointed if you seem to have no definite results of any sort. This is quite normal and the period of time it takes to become aware of positive results in the early stages varies from person to person. There is no correct time you should have to practise before you are ready to progress to the next stage of your development under the guidance of an experienced medium.

1. The right place and time

You need to select both a place and period when you are not likely to be disturbed. If there is a particular time of day when you can be alone this would be ideal. If not, then you must set aside a room where you can close the door on the outside world and know that your privacy will not be invaded. If possible, disconnect the telephone — its ringing would not do you any harm at this stage but you do not want to be disturbed in your practice.

2. Ask for help

Once you have found your quiet place and your quiet time and have settled yourself in your chair with your crystal, you must ask for help and guidance from the world of spirit itself. No medium would ever dream of beginning his or her work without asking for that guidance as well as protection. The words used for that protector may vary but that is of little importance. It does not matter whether you think of it as 'Spirit', 'Father', 'God', or any other word which may symbolize to you that ever-loving force which will guide you on your path. It is acknowledgement of that spiritual force and of your need for its help which is essential. The greatest mediums, just like the greatest healers, are only too ready to recognize that they are but the tools of a divine spiritual teacher who has enabled them to develop sufficiently to become aware of messages from the world beyond.

3. Relaxation

Sit in an upright chair, holding your crystal in your left hand. Taking your time, practise a relaxation technique. Starting with your feet, tense and relax each set of muscles in turn, working upwards through your legs, your thighs and your buttocks right up the whole of the trunk of your body. At each stage make sure that you can actually feel the contrast between the tension and relaxation of your muscles. Then turn your attention to your hands, your arms and your shoulders, doing exactly the same thing. Now you come to the most important — and the most difficult — area. Spend plenty of time tensing and relaxing the muscles of your neck and your jaw. It is in these areas and in the head itself that tension is most likely to build up during the course of our everyday lives. And it is in these areas that it is most important to dispel that tension — after all, it is the higher chakras which are the ones involved in your spiritual development and in receiving those thought forces which come to you from a higher plane. As time goes on you will find it easier and easier to relax — and this will benefit you in your ordinary life as well as in your development.

4. Opening the chakras

Using the method set out in the earlier chapter, and starting with the root chakra, concentrate on each of your chakras in turn. Do not hurry this process, particularly when you reach the brow and crown chakras; after all, you are opening the door to the world of spirit and you want to be certain that the door is indeed fully open.

5. Meditation

Begin by visualizing a stream of energy and light between the crystal you are holding and your third eye. Hold this image in your mind for as long as you can but when your mind starts to wander and other thoughts appear, allow this to happen. These thoughts, sensations and images — which may appear haphazard at first — are the beginnings of an increased awareness. There are those who would advocate 'making your mind blank' but this is virtually impossible for anyone other than those who are already highly developed and will only lead to disappointment for those who try and — as they will — fail to achieve this 'blankness'. Do not try to guide the thoughts which enter your consciousness in any particular direction, but try to be an 'observer', to note what occurs without imposing a logical explanation on it. It does not matter if you are not able at first to understand the images and impressions which come to you. Some may have little or no relevance and others may take time to make their meaning apparent to you.

6. Stop when you are tired

The concentration required in this process can be quite exhausting if maintained for excessively long periods. Never force yourself to continue once you begin to feel tired. It is not the length of time you are able to continue at one session which is important; it is the regularity of the sessions which is vital. If you wanted to run in a marathon, you would not go for one long jog and think that you were fit. You would practise regularly, increasing your strength and stamina little by little. During the meditation period you will be learning to use your 'spiritual' rather than your physical muscles, but the same comments apply. In the beginning you may certainly find that a period of ten minutes or so of this intense concentration is quite sufficient and there is no shame in this. In any case, should you allow yourself to become tired, you would not be able to receive any valid impressions and so you would be wasting your time as well as exhausting yourself.

7. Closing the chakras

Closing down the chakras is just as important as opening them in the first place. You do not want to leave yourself open and vulnerable to energy forces which you may not wish to receive. So, no matter how tired you may be, make sure that you always spend a little time on this process. Always begin with the crown chakra and work downwards to the root or base chakra.

8. Giving thanks

Before you began, you took some time to ask for help and guidance in your progress along the path towards spiritual awareness. It is just as important to give thanks for that help and guidance once you have completed your practice. Do not neglect this essential part of the process as it shows humbleness of spirit and integrity of purpose, both of which are vital ingredients of a medium.

9. Make notes

After each practice session, remember to make notes of what you have seen, heard or experienced. Even if the images appear to you to be vague and unconnected, they may in fact have a deep significance which may not make itself apparent to you until some time later. Never think that you will remember everything without making notes — you will not. And by remembering something inaccurately you may actually lose sight of its relevance.

Following these nine steps will certainly help you on your way if you decide that mediumship is the direction in which you want to go. But I would emphasize once again that it is only the beginning of the journey and that, once you realize that your awareness is increasing, you would be well advised to seek the assistance of an experienced medium who will be able to help you on a more personal level. As well as guiding you further along the path of development, a medium should be able to assist you in your understanding of the relevance of any symbolic images which may occur.

6.

Crystals and Astrology

Astrology is both a science and an art. Whereas at one time the validity and truth of astrology were victims of disbelief and even mockery, more and more people now accept its relevance in our lives.

Just as each sign in astrology has been allotted its own particular colour and its own particular flower, so too is there a specific gemstone linked by experts to that sign. Unfortunately, as there is more than one expert, you will find more than one crystal or stone corresponding to any particular sign. What I have attempted to do is to study the available lists and give details of those which are most commonly supposed to correspond to each sign.

Now if each sign of the zodiac has a specific stone which relates to it, it would seem to follow that those people born under a particular sign would do well to wear the appropriate stone. A birth chart, however, takes into account the position of each planet at the moment the individual was born, and each person is made up of characteristics which have been created by a combination of all those planetary positions. In the same way, you should not select your personal stone just because you were born at a particular time on a particular day. You should study the typical characteristics of each sign, decide which seems closest to your own personality and choose the stone which is linked to that sign. In addition, remember that none of our personalities remain static with the passing of time and you may well find that, as your temperament and nature undergo changes at different times in your life, you will feel happier wearing a different stone.

On the following six pages of this chapter are some personality charts which you may care to complete and which should give you some guidance as to your character at this particular stage of your life. Study all of them and place a tick against those items which seem to you to be the most apt descriptions of your nature. Then look at all the charts and see which one has the highest number of ticks. If you then refer to the list which follows, you will see which sign's characteristics are most like yours at the present time and also which gemstone or crystal corresponds to that particular sign. If you find that more than one sign appears to be applicable, it merely means that you will have more stones from which to choose to enhance your present personality.

CHART A

Positive characteristics:		
Enterprising		✓
Energetic	✓	✓
Freedom-loving	✓	✓
Enthusiastic	✓	✓
Adventurous	✓	✓
Direct in dealings with others	✓	✓
Negative characteristics:		
Impatient	✓	✓
Impulsive	✓	✓
Lacking subtlety		
Quick-tempered	✓	✓
Selfish		
Restless		✓
Total:	8	10

CHART B

Positive characteristics:		
Patient		
Reliable	✓	✓
Practical		✓
Warm-hearted	✓	✓
Good at business		✓
Determined		✓
Love of good things in life	✓	✓
Negative characteristics:		
Stubborn		✓
Possessive	✓	
Pompous		
Self-indulgent	✓	✓
Greedy		
Total:	5	8

CHART C

Positive characteristics:	
Versatile	✓
Intellectual	✓
Good communicator	✓
Witty	✓
Adaptable	✓
Logical	✓
Youthful appearance	✓
Negative characteristics:	
Restless	
Inconsistent	✓
Lives on nerves	
Changeable	✓
A gossip	
Total:	9

CHART D

Positive characteristics:	
Sensitive	✓
Kind	✓
Thrifty	
Intelligent	✓
Cautious	
Good imagination	✓
Strong sense of family	✓
Negative characteristics:	
Hyper-sensitive	
Quick-tempered	✓
Unforgiving	
Untidy	✓
Over-emotional	✓
Total:	8

CHART E

Positive characteristics:	
Generous	✓
Creative	✓
Broad-minded	✓
Sense of drama	✓
Magnanimous	✓
Enthusiastic	✓
Negative characteristics:	
Pompous	
Intolerant	✓
Conceited	✓
Excessive love of power	✓
Dogmatic	
Snobbish	
Total:	8 6

CHART F

Positive characteristics:	
Tidy	
Modest	✓
Meticulous	✓
Analytical	✓
Hard worker	✓
Eager to help	✓
Charming	✓
Negative characteristics:	
Worrier	✓
Fussy	✓
Over-critical	✓
Difficulty in relaxing	✓
Stand-offish	
Total:	5 10

CHART G

Positive characteristics:	
Charming	✓
Romantic	✓
Love of beauty	✓
Idealistic	✓
Easy-going	✓
Generous	✓
Negative characteristics:	
Frivolous	✓
Too easily influenced	✓
Indecisive	✓
Gullible	✓
Resentful	
Flirtatious	✓
Total:	11

CHART H

Positive characteristics:	
Strong emotions	✓
Passionate	✓
Powerful	✓
Highly imaginative	✓
Attractive to others	✓
Determined	✓
Negative characteristics:	
Jealous	
Stubborn	✓
Secretive	
Suspicious	✓
Intractable	
Self-indulgent	✓
Total:	9

CHART I

Positive characteristics:	
Optimistic	✓
Versatile	✓
Good-humoured	✓
Sincere	✓
Dependable	✓
Philosophical	✓
Negative characteristics:	
Tactless	
Careless	
Irresponsible	
Tends to exaggerate	✓
Extremist	✓
Restless	✓
Total:	9

CHART J

Positive characteristics:	
Reliable	✓
Sensible	✓
Patient	✓
Persevering	✓
Careful	✓
Ambitious	✓
Prudent	
Negative characteristics:	
Pessimistic	
Miserly	
Obsessed with conventionality	
Rigid in outlook	
Tendency towards depression	✓
Total:	7

The Crystal Workbook

CHART K

Positive characteristics:	
Friendly	✓
Faithful	✓
Idealistic	✓
Intelligent	✓
Humanitarian	✓
Independent	✓
Original in thought	✓
Negative characteristics:	
Rebellious	✓
Tactless	
Eccentric	✓
Perverse	
Contrary	✓
Total:	9 / 10

CHART L

Positive characteristics:	
Compassionate	✓
Sympathetic	✓
Kind	✓
Intuitive	✓
Sensitive	✓
Desire to help	✓
Negative characteristics:	
Vague	
Easily confused	
Weak-willed	✓
Secretive	
Unable to cope with life	
Indecisive	
Total:	7 / 9

Crystals and Astrology

Interpretations

Once you have added up the number of ticks in each of the personality charts, it should become apparent to you which chart is most applicable to you at this time in your life. It could, of course, turn out that two or more charts display the same number of ticks — and really this should not be too surprising as each of us is a character with many different aspects to our personality. In this case you will merely have a large number of stones from which to choose. Remember that it is your present personality which is important, not your birth sign — although naturally in some cases these two will tally.

Chart A	You have a strongly Arian character and the most appropriate stone for you would be the diamond, the ruby or — rather less expensively — the red jasper.
Chart B	You are typical of the warm-hearted Taurus character. If you are a Taurus-type, you should choose as your personal stone the sapphire or the lapis lazuli.
Chart C	Yours is the lively communicative nature of the Gemini-type and the most suitable stone for you would be the citrine or one of the yellow agates.
Chart D	The sensitive, vulnerable personality of the Cancerian subject is most similar to yours at this period of your life and the ideal stone for you would be the pearl or the moonstone.
Chart E	The characteristics featured in this chart are typical of the personality of those born under Leo. If you are passing through a Leonine stage in your life, the stone for you to choose would be the tiger's eye.
Chart F	At this moment in time your nature is that of the hard-working Virgo-type. In this case the stone for you to wear or carry would be the green jasper or the sardonyx.
Chart G	You have the charming, if somewhat indecisive, nature of the Libran subject. To enhance your personality you would do well to choose the sapphire as your personal gemstone.
Chart H	Your personality at present most closely resembles that of the passionate Scorpio-subject. If you are a Scorpio-type the most appropriate stone for you would be the ruby, the opal or the red jasper.
Chart I	You display the characteristics of the freedom-loving Sagittarian and the most suitable stone for you would be the topaz.

Chart J Determination and conventionality are typical of the Capricorn-type and would appear to be facets of your personality at the present time. If this is the case, you should choose as your stone the turquoise or the smokey quartz.

Chart K At present your nature is that of the delightfully unpredictable Aquarian and the gemstone most appropriate for you during this phase of your life is the amethyst.

Chart L You are as sensitive and as emotional as those whose birth sign is Pisces. Like all those with a Piscean nature you should add balance to your life by wearing a moonstone.

7.

Divination

Using crystals and semiprecious stones as tools of divination is nothing new. In fact — possibly with the exception of sand — this form of reading, known officially as *lithomancy*, is probably one of the oldest known to man. It may well have been used in the lost continent of Atlantis; after all, the inhabitants of that ill-fated land were certainly well aware of the value of crystals as a means of healing and of providing a source of energy. Why should they not have been equally aware of the value of crystals in giving pointers towards the future of an individual or even of a race? Crystals and gemstones were certainly used in this way in civilizations as ancient as the Egyptians, the Aztecs and the Incas — and it was because I had been reading about these particular civilizations that I first began to use crystals as my own divinatory tools.

It was one of those 'coincidences' of life (although, like Jung, I do not believe that coincidences actually exist — I think that opportunities are put before us and we either take advantage of these opportunities or we do not). I had been a Tarot card reader for some years and at the time was quite happy with this way of working. Then, in 1981, I was given as a present a box of about 150 beautifully polished crystals and gemstones. I was not given them for any reason other than because they were beautiful and the donor believed that they would appeal to me because of that beauty — which, of course, they did. But, as I examined these lovely jewels of nature, I was aware of different sensations emanating from the different types of stone. Some appeared to have a strong pulsating energy all of their own, while others felt calm and peaceful. Some felt hot and almost 'alive' as I handled them, while others did not seem to be emitting any sensation at all.

As I examined my present, I began to recall details from a book which I had been reading — a book about the civilizations of the Aztecs and the Incas. In that book there had been one whole section devoted to the intuitive abilities of those people and I had read that they actually used crystals and gemstones as their means of divination. By using them they had been able to answer questions and to predict the future.

As I have said, I do not believe in coincidence and so I decided that there had to be a reason for the synchronicity of the gift and the book I had been reading. The answer seemed obvious; I was meant to use my own crystals and stones as my tools of divination, just as they had been used all those centuries before. Soon I would be able to help those who were to come to me with questions about their future by using my own collection of crystals and stones.

One of the things which added to the pleasure of using gemstones in this way was that they are things of nature rather than man-made items. They had been polished, but this merely brought out the natural beauty of their shapes and colours. In addition there was a great deal of pleasure to be gained merely by handling them, and this would only add to the satisfaction of my work.

But, although the book I had been reading had told me that crystals and stones had been used in divination by the Aztecs and the Incas, it had not told me *how* they had used them. I knew that each individual stone had its own meaning within the context of a psychic reading — but who was there to explain to me just which meaning was to be attributed to each of *my* jewels of nature? There was only one answer. I had to meditate upon them, one by one, and ask those who guide me to make the interpretations clear to me. And that is just what I did. Little by little, over a period of several weeks, the meanings were given to me. Those meanings I shall explain to you later in this chapter.

Now we have to consider what is meant by divination as opposed to 'fortune-telling'. The latter always seems to carry with it the impression of the gypsy at the end of a seaside pier who will tell you about a 'tall, dark stranger'. The value of a sincere psychic reading, whatever method is used, is that it can give a sense of direction and of hope to someone who may feel that he or she has lost his way in life. It is never the task of the reader to make decisions for the questioner — that must be left to him alone. Hopefully the reader will be able to point out the various choices which may lie ahead and any areas where particular care must be taken or where it seems that the situation would be most advantageous. Having done that, it is up to the questioner himself to decide what he is going to do with the opportunities which confront him.

Whether one is dealing with crystals, Tarot cards or any other means of divination, there are two stages to the reading. Naturally, as each stone or card has a meaning, the reader will have to study the ones chosen by the questioner before beginning. Ideally, however, a good psychic reader will employ both his knowledge and understanding of his tools and also his intuition or psychic ability. Hopefully, by the time you start to consider reading from the crystals and gemstones, you will already have worked to develop your own psychic ability as mentioned in the earlier chapter. By using that ability, in conjunction with the knowledge you are about to acquire as to the meanings of the individual stones, you should be able to give a sincere and helpful reading to any who may seek your help.

Acquiring Your Stones

This is the only instance where I would suggest that crystals and gemstones can be acquired by post or bought for you by someone else. As we have already seen, when you are to use crystals for healing, meditation or development purposes, it is the individual stone and your affinity with it which is of vital importance. In a reading, however, it is somewhat different. In this case every moss agate will mean the same as every other moss agate, so it would be sufficient for you to say that you needed one, two, or more moss agates. You do not have to select the actual ones. In addition, as your stones will be handled by many other people rather than by you alone, you do not need to own 'personal' stones at all.

There is no correct number of crystals and stones needed in order to be able to read from them. I have about 150 — but there are no more than 50 or 60 different types of stone in that collection. Almost all of them are repeated — some of them several times. I find that this is often helpful in that, should a questioner select several stones of the same type, it usually indicates that there is one particular area on which we need to concentrate. Also, having several stones of the same type can help to give an idea of time. For example, I have four amethystine agates in my own collection. Each of these lovely grey and pink stones would refer to a move of home, but I have found that the larger the stone chosen, the nearer in time that move would appear to be. So it can only be advantageous to have a greater number of stones than the bare essentials which you will find listed further on in this chapter.

Once you have acquired your stones, you must wash them in clear running water and then dry them. You will probably find that they will need washing fairly frequently as they will be handled often by many different people and will soon begin to lose their lustre if dirty. Since one of the great joys of working with crystals and gemstones is the pleasure one can receive from looking at them and touching them, it is only sensible to keep them clean and bright.

What Else Will You Need?

You will need some sort of dish or tray in which to display your gems. I have an oblong tray approximately 12 inches by 9 inches, lined with black velvet. A flat tray is better than a round dish as it enables the prospective questioner to see the stones more easily and to move them around in order to make his selection.

You will also need some sort of mat or cloth for the stones to be placed upon once they have been selected. Mine is made of black velvet and this material was chosen very deliberately — you have only to look in the window of any jeweller's shop to see that dark velvet is almost always chosen as the perfect foil to the beautiful articles on display. Although we may not be using perfectly cut diamonds and other precious stones, do our own gems not deserve the same setting so that their particular beauty may be shown off to the greatest advantage? It is not essential for you to use black velvet to set off your gemstones — you can use any fabric, or even paper or card. But if you try putting your own crystals on a piece of black velvet, I think that you, too, will feel that no other material does your stones sufficient justice.

For another method of laying out the stones, you will need a large circle, divided into 12 equal segments (relating to the 12 astrological houses). This can be drawn on paper or card or embroidered on cloth.

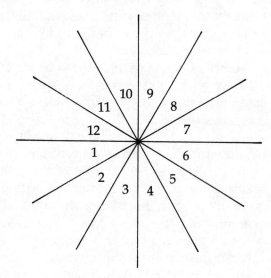

Figure 14

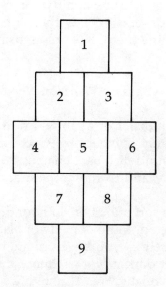

Figure 15

For the third of the divinatory methods I am going to tell you about, you will need a larger version of Figure 15. Once again this may be drawn on card or embroidered on a piece of cloth.

You will notice that I have mentioned three different techniques for conducting a reading from crystals and semiprecious stones. These are the three that I use and which I have found to be both accurate and effective. You may use these methods for yourself — indeed they will make an excellent starting-point. But always remember that you are an individual and in theory there are as many different methods and layouts as there are readers wishing to use them. If you intuitively feel that a different method would be most appropriate for you, then by all means go ahead and use it.

Meanings of Crystals and Stones in a Reading

The list which follows contains the 32 stones which I consider to be essential before you can even begin to give a balanced reading. Remember, however, that this is a minimum number and you can have as many of each variety of stone as you wish. Indeed your task can only be made easier if you have more than one of each, as you will see.

Inside the front and back covers of this book you will find colour illustrations of all the gemstones contained in the list. You would find it easier to understand what follows if you spend a little time learning to recognize all the stones (especially those which you do not already know) and comparing them with their meanings as described in the list.

Agate geode

The agate geode represents the potential psychic ability of the questioner. Some people choose to place this stone on the mat with its glittering hollow on display; others will turn it upside down and place it face downwards so that only the rough exterior of the stone is visible. This is very significant. If the questioner places the stone cut side uppermost so that you can see the sparkling lilac or mauve interior of the hollow, this would mean that he is either using his own psychic ability or at least is sufficiently aware of it to be developing it. Should the agate geode be placed face downwards on the mat, however, this would mean that that intuitive aspect of the questioner's nature is being blocked. In some cases those who are frightened of their own psychic ability will block it quite deliberately. In other cases, however, it may be that outside circumstances are imposing a block upon the questioner. Suppose he were in a state of depression — or perhaps had recently been ill or suffered a bereavement. In any of those eventualities his psychic ability would be far less likely to function well.

Sometimes the questioner will turn the agate geode over and over in his hand before deciding which way to place it on the mat. In many cases this will indicate that either his own psychic powers are just beginning to make themselves felt or that he is not at all sure of himself and his own ability in this particular sphere.

You will see, therefore, that it is essential to watch closely as the stones are selected — and indeed how they are selected. It is not sufficient merely to look at them once they have been placed on the mat. It is just as important to observe what happens as they are chosen. It is possible to learn a great deal about the questioner from those crystals which he rejects or which he handles for several moments before finally selecting. Your task begins as soon as he starts looking at the crystals and stones as they lie in the tray.

Agate quartz

The agate quartz represents a young man — often a son. It is one of those crystals whose meaning is strongly affected by those stones which surround it. If the stone closest to it happens to be the dark green jasper, for example — the stone which represents worry or anxiety — it is possible that the questioner is feeling concerned about his son for some specific reason. It could be that the young man's health has been less than perfect; it could be that he has difficulty in finding a job — or it could merely be that he is awaiting the results of examinations and is feeling rather anxious. Whatever is the cause for the concern, you have merely to take note of the other stones and crystals which surround the agate quartz to glean information as to the outcome.

It is possible that this crystal will refer to a young man who is not actually the son of the questioner but it will always be someone for whom he cares, perhaps a relative or someone whom he has befriended. The agate quartz in a reading will never refer merely to a casual acquaintance. If you find that the agate quartz appears to be isolated in the reading, you may find it helpful to ask the questioner to select a further three stones in order to give you further information.

Agate with fossils

Although this stone is often quite dull in appearance, being brown with creamy-coloured fossils embedded in it, its presence in a reading is always a good sign. The agate with fossils always indicates an increase in money for the questioner. If you look at the stones which surround it you will often see the reason for this increase. If there is a turritella agate nearby, then it is likely that a change of job will bring greater reward; if you see the Botswana agate it could be that the questioner will either receive a gift or perhaps win some money. So, plain as the stone may look, any questioner should be pleased when he has been led to choose it.

Amethyst

We have already seen how the amethyst can be used in spiritual healing and in meditation. The lovely mauve amethyst is always the spiritual crystal. Its presence in a reading often indicates an interest in spiritual philosophies and in development. One will often find that there may be several amethysts chosen in one reading and this is not difficult to understand when you consider that, if someone has strong spiritual beliefs, they are likely to affect his life in many different ways.

Once again it is necessary to take note of the surrounding stones. If, as is quite common, you will find a rose quartz in close proximity, it may be that the questioner is drawn towards spiritual healing; if you see the glittering agate geode, he could well have clairvoyant or psychic tendencies.

As well as the amethyst itself, there is also an eight-sided amethystine octahedron. Although this crystal too indicates spiritual beliefs, its presence in a reading is not quite such a good sign. The amethystine octahedron will often be chosen by someone who becomes so deeply involved in what he considers to be 'spiritual' that he is unable to cope with ordinary day-to-day living. It is usually selected by the type of person who blindly follows the teachings of some distant guru or who refuses to take charge of his own future. He is also likely to have difficulties coping with human relationships as he sees the world through the unrealistic eyes of the supreme idealist.

Amethystine agate

It is often as well to have more than one amethystine agate in your collection. I have four in my own set of stones; each has the same unmistakable grey and pink markings but each is a different size.

The amethystine agate always indicates a move of home and the reason for suggesting that it would be advantageous to have more than one in your collection is that the size of the stone will often give an indication of how far away in time that move is to be. The larger the stone the sooner the move is likely to arise. If the amethystine agate is placed on the mat near to a turritella agate, it is possible that a change of job will be the reason for the move. If the stone nearby is the irridescent labradorite, it is quite likely that the move will be to another country.

I usually give readings which cover approximately two years from the date of the consultation. The largest amethystine agate, therefore, would indicate the likelihood of a move in the first six months of that time and the smallest in the last six months. The two other stones would relate to the interim period.

Aquamarine

The aquamarine is long and cool in appearance and this should help you to remember its meaning in a reading. It represents clear thinking, logic and a common-sense approach. It can also indicate that the questioner who has come to consult you is unsure of the validity of any form of psychic reading and finds it difficult to believe anything for which he cannot find a logical explanation. He will probably believe those parts he chooses to believe, rejecting the rest completely. That, however, is his problem and not yours. As long as you know that you have given as honest and accurate a reading as you can, you cannot take responsibility for the way in which any questioner responds to that reading. Indeed you will often find someone who has been sceptical and even critical at the time of the reading will contact you some time later to tell you how accurate you have been and to apologize for his previous attitude.

If you see a number of crystals and stones indicating agitation and unease in the life of the questioner, the presence of an aquamarine would be a welcome sight as it would indicate that the questioner has only to stop and think and allow his logical, common-sense side to come to the fore in order to make order out of disaster.

Bloodstone

The dark green stone with its flecks of dark red 'blood' will usually indicate physical problems of some sort — anything from toothache upwards. It is never, however, a 'doom and gloom' sign; you will always find that wherever a negative stone appears in a reading it is there to serve as a reminder to the questioner to take care rather than to be a sign of impending disaster.

The bloodstone does not necessarily mean that either the questioner or anyone close to him has anything drastically wrong. Indeed, depending upon its position in a reading, it could be an indication of a past problem, one from which the questioner has already emerged satisfactorily.

Sometimes the bloodstone will appear in a reading in close juxtaposition to a stone which represents another person, and in this case it is likely that the physical problem concerns that person rather than the questioner himself.

It is important when dealing with this or any other stone which could indicate problems or anxieties to send the questioner away in a positive frame of mind. As we have already seen, it is your task to be helpful and to give hope and encouragement as to the future, while at the same time detailing honestly what you see.

Quite often you will find that, although the presence of the bloodstone indicates that there is some pain or discomfort which manifests itself in a physical way, the cause of that pain or discomfort could well be stress or tension on the part of the questioner and all that is needed is a little advice on how to cope with this. That advice you will be able to give, having already read the sections on relaxation and meditation.

Blue lace agate

We have seen how the agate quartz refers to a son or a young man. In just the same way, the delicate blue lace agate represents a daughter or a young girl. Once again you must concentrate on the stones which surround to define the reason for its presence in a reading. Perhaps there has been some sort of friction or some disappointment between the questioner and the young girl concerned; in this case you should be able to give some indication of the outcome of this situation. Perhaps there is a particular reason for joy and pleasure concerning the young girl and the reason for this, too, should become apparent when you look both at the surrounding stones and the position of the blue lace agate in the reading.

Botswana agate

The lovely Botswana agate, looking for all the world like an old-fashioned 'bull's eye' indicates that an unexpected delight lies ahead for the questioner. The meaning of this stone can, of course, vary considerably — ranging from the trivial to the vitally important. For one of my clients, who had been adopted when she was just a few weeks old, it represented a reunion with a sister she had not seen and whom, until she was nearly 40, she did not know even existed. For another it was a win (albeit not an enormous one) on the premium bonds.

You will have to consider the surrounding stones and the position of the Botswana agate itself in order to decide upon the area of delight to which it refers.

If someone has come to you with a specific problem on his mind, the presence of the Botswana agate in his reading will usually indicate a happy solution to that problem.

Citrine quartz

The citrine quartz is a fusion of a beautiful amber-coloured citrine and a frosty quartz crystal — and the effect is quite delightful. Its meaning in a reading is 'new beginnings'. The new beginnings can relate to a definite event, such as a new job, a new interest or a new relationship. It can also indicate that the questioner is entering a new phase of his life.

The new beginning indicated by the presence of the citrine quartz tends to be one which has 'appeared' in the life of the questioner as opposed to one which he has been actively seeking. Whether it is a new relationship or a new job, the likelihood would be that it would appear to come about 'by coincidence' or without effort on his part — something, in other words, which has been handed to him on a plate.

Dendritic agate

The meaning of the dendritic agate is 'sunshine' — and this can be taken in the literal or symbolic sense. It can refer to a place where the sun often shines; if it appears close to a piece of labradorite (meaning overseas travel), it could refer to a sunny rather than a chilly spot. It can also mean that the timing of a particular event is to be in the 'sunshine' time of the year.

The symbolic meaning of the dendritic agate is that the outcome of a particular situation will be 'sunny' — in other words, successful and pleasurable. Perhaps it marks the end of a difficult and stressful period in the life of the questioner.

If your questioner asks a specific question requiring the answer 'yes' or 'no', and the dendritic agate is present in the reading, then the answer is almost certain to be in the affirmative.

Fluorite octahedron

The presence of the fluorite octahedron in a reading would indicate artistic or creative abilities on the part of the questioner — and these would be as a practitioner (although not necessarily professional) rather than as an onlooker. Of course 'artistic' does not necessarily refer to painting and drawing; it could just as easily be, for example, writing, designing or a talent for music.

The fluorite octahedron has both a positive and a negative aspect to it. Its positive side is that it would indicate quite a high degree of talent on the part of the questioner in whichever creative field appeals to him. He would be likely to gain pleasure and achieve success should he pursue this talent. The negative side of the crystal is that the questioner may expect such high results from his creative abilities that he is never satisfied with anything he is able to achieve and this can lead to discontent or even, in more extreme cases, to the refusal to attempt anything on a creative level because of the doubt he feels in his own talent.

If you feel that the latter case is the true one, then this is the time for a little gentle advice to the questioner. The type of person who refuses to try for fear of failure in any one aspect of his life is likely to carry that tendency over into other areas of his life too, and the result could be a sad and introverted person who eventually tries nothing at all in case he does not succeed.

Green jasper

The presence of any jasper in a reading relates to the emotions and feelings of the questioner. The red, light green and variegated jaspers will be dealt with further on in this list. The green jasper — and by this is meant the dark green jasper which is almost the colour of the bloodstone but without those distinguishing red spots — is not the happiest of stones to find in a reading. It indicates that the questioner feels that he is unloved, unsuccessful — even unworthy of love. Of course, it does not necessarily mean that this is actually so, merely that the questioner feels that it is. You will have to decide which is the case in each reading that you do. Perhaps he merely fears that he will be rejected emotionally and so is hesitant about making his own feelings clear; perhaps he thinks that he would not stand a chance of being successful and so he refuses even to make an attempt. A dark green jasper in a reading certainly means that advice and help is called for — after all, emotions are just as deeply felt, even if they are based on a false premise.

Iron pyrites

Another name for iron pyrites is 'Fool's Gold' and the stone gained the name because its golden gleaming cubic form used to fool the prospectors for gold into believing that they had found the real thing. In a reading the iron pyrites is a stone of which to take particular note. It indicates deception, mistrust — something which is not what it seems. It never refers to something trivial but is always of significance in any reading.

By examining the stones which lie around the iron pyrites, you will be able to see where that deception lies. It could be that the questioner is being deceived by a friend, a loved one or a business acquaintance. It could be a sign that a step which he is contemplating should not be taken without very careful consideration. It could also be read in more general terms where the meaning would be that the questioner is perhaps someone who tends to place his trust a little too easily — particularly if he is someone who is by nature extremely trustworthy himself. It could even mean 'blind faith' in the sense that the questioner might be inclined to think that 'everything will work out all right' without putting in sufficient effort to ensure that it does. While you do not wish to encourage abject pessimism in the person for whom you are conducting the reading, none the less it might be advisable in such cases to give a little timely advice that care should be taken in a specific area.

All this does not mean that the presence of the iron pyrites in a reading is necessarily a bad sign. On the contrary, if it is sufficient to make the questioner stop and think before rushing headlong into some unsuitable commitment, then it is actually doing him a favour and should be regarded as such.

Labradorite

The incandescent quality of the beautiful labradorite is reminiscent of butterflies' wings. The stone itself in a reading indicates a connection with a country overseas. Once again you have to look at the position of the stone and the stones which surround it. It could be that the questioner is contemplating making a business trip, visiting relatives who live abroad or even — particularly if the red jasper of emotion is close by — about to begin a relationship with a person from another country.

It is not usual for the labradorite merely to represent a summer holiday — its meaning is normally more significant than that. Bearing that in mind, it would be possible for it to mean a vacation if there were to be some specific and relevant event taking place. It could even be that, by telling the questioner that a trip overseas is likely, you would in some way be answering some other unspoken question — such as a concern to do with financial status.

Light green jasper

A delightful stone to find in any reading, the clear golden green of the light green jasper almost shouts its meaning. It is joy, and happiness. This stone — one of the few jaspers to contain not a hint of red — does not represent the tranquillity of contentment; it bubbles over with exuberant delight.

The light green jasper is most usually selected when a particular question has been asked — and it is always to be taken as a sign that the questioner will be extremely happy with the answer.

Moss agate

After the energetic joy of the light green jasper, we come to the more gentle contentment depicted by the moss agate. A stone which is always a pleasure to find in any reading, the moss agate often signifies an end to a period of anxiety or stress. On some occasions a particular questioner will select several moss agates in the course of his reading and this could mean that he is trying so hard to achieve peace of mind that he is failing to see that it is actually within his grasp.

If someone has been to consult me on several occasions because of difficulties which face him in his life and, on a subsequent reading, he selects a moss agate as one of his chosen stones, I always take this as a sign that things are beginning to improve and that he has more to look forward to than he previously had.

Petrified wood

The petrified wood relates to law and legal matters. Of course these do not necessarily have to be troublesome matters — we could be talking about the gaining of a new contract for an ambitious businessman. Naturally the meaning could be less fortuitous too — perhaps a divorce is in the air or a dispute over a will. Look at the position of the piece of petrified wood and the stones which surround it to discover its true meaning in the particular reading. For example, should the piece of petrified wood be placed next to the iron pyrites it could be that there is some legal document which needs to be more closely examined, possibly by an expert.

Pink and grey jasper

You have seen how the blue lace agate refers to a young girl or daughter and the agate quartz refers to a young man or son. The pretty pink and grey jasper is one of these 'person' stones and it signifies the presence of an older person — man or woman. This would never mean someone who is just a year or two older than the questioner, but usually a generation or more his senior — and quite often a parent or someone who has stood in the place of a parent. The presence of the pink and grey jasper in the reading would indicate that a situation concerning this older person is going to affect the questioner's life in some way. Perhaps he is to receive a gift or a legacy from this older person; perhaps the lady or gentleman will be coming to live with him. Whatever is the case, it is bound to refer to something which will affect the future of the questioner and his hopes and plans.

Purple agate

The purple agate is one of those stones which has two meanings in a reading. The first is quite simple; it represents water. Look at its position and the stones which surround it. Is the questioner about to move to a place which is near water — whether the sea, a river, or a lake? It is unlikely to indicate a move *overseas* unless a labradorite is found nearby.

The other meaning of the purple agate is sensitivity. This can be an excellent thing if we are dealing with someone who is creative by nature or who is involved in some healing or caring profession. The other side of the 'sensitivity coin' however could indicate someone who is so over-sensitive that it affects both his life and his judgement. If the purple agate appears among stones which indicate business or legal matters, it may be there as a gentle warning to the questioner that he needs to come down to earth a little and deal with things on a more practical level.

Quartz crystal

In earlier chapters you have seen how effective the quartz crystal is in relation to meditation, healing, etc. Later in this book you will learn about its efficacy in relation to dowsing and other techniques. But for the moment we are merely concerned with its meaning when it is selected in the course of a psychic reading.

The quartz crystal indicates energy, vitality and strength. If we are dealing with someone who has been unwell, the presence of the quartz crystal would indicate a return to robust and vibrant health. Equally it can refer to the questioner finding the inner energy to take charge of his own life or the strength needed to cope with a particular situation.

It is always a positive sign when the quartz crystal appears in a reading, as it means not only that the questioner will be successful but that this will be because of his own achievement rather than because of what has been done for him.

Red jasper

As you will have seen in those we have already considered, all the jaspers refer to deeply-felt emotions — some positive and some negative. The glowing red jasper can indicate any strong and energetic emotion such as love, jealousy or even anger. If you find a large number of red jaspers in one reading, that will be a certain sign that the questioner's emotions are to the fore — although it is possible that this may be on a subconscious rather than a conscious level.

The red jasper will always relate to strong emotions concerning personal relationships rather than to facts. We are not talking here about the relationship between only one man and one woman: it could be the love between parent and child, the love between close friends — even the love felt by the questioner for mankind in general.

Rose quartz

The lovely rose quartz is the healer's crystal. It would indicate that the questioner has healing abilities — whether or not he recognizes them and whether or not he chooses to make use of them. Do not forget that there are many forms of healing. Spiritual healing is, of course, one type — and if this is what is meant you would be likely to find an amethyst close by. But a nurse is a healer, as is a veterinarian. If you find an aquamarine in close proximity to the rose quartz, the healing may well be of minds rather than bodies. Perhaps the questioner has the makings of a counsellor or therapist.

Sometimes the rose quartz will not appear in the original reading but only when the questioner wishes to ask a question about a specific person. If this is the case, it usually means that that person is in need of healing — whether that healing is mental, physical, emotional or spritual.

Ruby

Perfectionism — that is the meaning of a ruby in a reading. Of course perfectionism can be an excellent attribute, indicating hard work and a desire to give of one's best. However, it can also mean that one is too hard on oneself or others, expecting more than can possibly be achieved. In some cases it can indicate that the questioner is such a perfectionist and so afraid of failure that, rather than do something less than perfectly, he prefers to do nothing at all.

Of course, if you are giving a reading to a student who is awaiting the results of an examination, the presence of the ruby in the reading is something with which that student should be delighted as it indicates the attainment of a high standard. However, if the questioner is a concerned parent, it is possible that he is expecting too much of his son or daughter — more than he or she is able to achieve — and could thereby be causing the young person unnecessary stress or tension.

In some extreme cases the presence of the ruby can indicate a preoccupation with detail which is similar to obsession; in such a case it should be taken as a warning to the questioner to be a little more realistic in his aims.

Rutilated quartz

This is a lovely pale golden crystal which has threads of what look like pure gold running through it. In a reading it refers to artistic ability or creativity on the part of the questioner. Sometimes that creativity will not have been recognized or acknowledged; on other occasions it will be a vital part of his life. If the rutilated quartz is placed close to the agate with fossils (which indicates money), it is quite likely that the questioner either does or will earn his living in some artistic way.

Sometimes, having selected the rutilated quartz, the questioner will insist that he has no artistic ability whatsoever. If that is the case, it is likely that the pursuit of some creative hobby or interest would, in fact, be very beneficial to him. Perhaps he is working too hard and spending so much energy in this way that he has forgotten how to get pleasure out of life. It could be that he finds it difficult to verbalize his emotions and a creative outlet would help to release the tension caused by this situation.

Serpentina

To find the true meaning of the serpentina within a reading it is necessary to consider its position and the stones which surround it. It is one of those qualifying stones which add to information already there. It can often be of assistance when one has to differentiate between two or more people or things. If, for instance, you wish to refer to a young woman, possibly the daughter of the questioner, it could be that he has two or more daughters. The presence of the serpentina will indicate that we are talking about the eldest of them.

In the same way, if it appears that the questioner is going to become romantically involved with someone he already knows and if he feels that there are two possibilities, the serpentina would make it probable that the relationship is likely to be with the person he has known for the greater length of time.

So in effect the serpentina's meaning can be said to be 'oldest' or 'most mature' — whether it refers to people or to things.

Another meaning of the serpentina is the summer period — from about early June to late September. This can be of assistance when trying to give details as to the timing of a specific event.

Tektite

Tektite is dull and black — looking rather like a cross between a small piece of coal and a shrunken prune. In a reading it indicates feelings of hopelessness and of despair and even a tendency to give up on life and withdraw into oneself. Of course this does not mean that it has to refer to the questioner himself — although it may well do so. Quite often it will occur in a reading as a means of pointing out to the questioner that someone he knows is in need of help and support.

If you find that the tektite is surrounded by crystals of a more vibrant and positive nature, such as the quartz crystal, it would appear that the questioner has actually been going though a period of depression but that he is beginning to overcome this and that there are definitely hopeful signs for the future.

You might be tempted to draw the conclusion that the presence of the tektite would always give a gloomy and disturbing tone to the reading, but I have not found this to be so. After all, whenever something less than happy occurs in our lives, it is in fact all part of a great learning process and can indeed make us stronger and more able to cope. A piece of tektite need not be an ominous sign at all but rather a means of giving the questioner advance warning of a situation so that he may be prepared for it — or even so that he can take action to avoid it altogether.

Tiger's eye

The tiger's eye can represent the positive or the negative side of being alone. It can mean independence, the ability to stand on one's own feet, the need to be one's own person. For a young adult to leave home and make his own way in life is a sign of the need for independence; for a business man to contemplate starting his own company can be a strong and positive move. The tiger's eye, however, can also mean being left alone — perhaps after the breakdown of a marriage. Although both these meanings are possible, the former is the more likely as the tiger's eye has very strong links with the astrological sign of Leo and a positive outcome is therefore far more likely. Sometimes the independence represented by the tiger's eye will have been thrust upon the questioner rather than being actively sought by him — but the outcome is usually beneficial in any case (although it may well be that he will find this difficult to understand at the time).

It is not unusual to find several pieces of tiger's eye in one reading and this would indicate that independence and self-fulfilment are of vital importance to the questioner and that much of his future will depend upon the way in which he deals with that independence when it confronts him.

Turquoise

To find a turquoise in a reading is to find feeling of tranquillity and peace. Whereas the light green jasper depicts a positive and effervescent mood of joy, the gentle turquoise is a more subtle — and often longer-lasting — feeling of contentment.

If the turquoise is selected when stones are chosen so that a specific question may be answered, it is an almost certain indication that the questioner will be content with the answer he receives.

Turritella agate

The turritella agate with its snake-skin-like markings indicate a change in the work situation of the questioner. It could mean something as simple as an increase in status or responsibility in the place where he is already working. It could equally refer to the more dramatic changes brought about by a move to another job altogether.

Look at the position of the turritella agate in the reading and the stones and crystals which surround it. If you see an agate with fossils nearby, it would seem apparent that the change is certainly to be of financial benefit. Perhaps there is a labradorite close by indicating either that the questioner is likely to work abroad or that the firm by whom he is going to be employed has foreign connections. A tiger's eye is near to the turritella agate, it could well mean that the questioner is going to work for himself — and should there also be a dendritic agate there, you will be able to tell him that there is much to be gained from this move in terms of personal satisfaction.

Variegated jasper

This jasper, which is made up of a fusion of red, yellow and green jaspers, represents anxiety or lack of confidence in affairs of the heart. It often means that the questioner has had experiences in the past which have left him feeling unsure of himself in this particular area of his life. It may even be that these past experiences have caused him to erect invisible barriers around himself in order to become less vulnerable emotionally and that all that these barriers are really doing is effectively keeping him away from the possibility of emotional contentment in the future. If the reading helps him to acknowledge that these barriers exist and that he wishes to do something about breaking them down so that he can find the love and happiness he seeks, then the presence of that multi-coloured jasper will have done him a favour indeed.

White and amber agate

The white agate surrounded by a border of amber concerns books and writing. It could be that the questioner is about to undertake some course of study — particularly if there is a piece of aquamarine close by. Of course it could also mean that he is the teacher! If you find a rutilated quartz — indications of creative ability — near the white and amber agate, it could well refer to creative writing (and this could imply writing music in addition to writing words) and, should the agate with fossils also be present, it is possible that the questioner is going to take up some form of creative writing as a means of earning his living.

Consider the position of this stone and look closely at the stones which surround it and remember that accountants write in books as do lawyers, so we are not necessarily concerned only with budding authors or lyricists.

You have now had placed before you a comprehensive list of those crystals and stones upon which a reading can be based. Naturally, there is nothing to prevent you from having more than one of each type of stone — in fact it would be advantageous — but it is possible to give a basic reading just from one each of the varieties listed here. All of them are readily available from the firms listed at the end of this book as well as from may specialist gemstone shops and societies.

Once you have acquired your own stones you must take some time to get to know them really well. Compare them to the illustrations inside the front and back covers of the book and go over them time and time again until you are able to recognize each one and to know instantly what it indicates within a reading. Although you will naturally be eager to practise giving readings yourself, you will find that, until you are really well acquainted with your collection of crystals and stones, you will not be able to give any detailed information to those who may seek it. Every Tarot card has a different meaning, every rune stone has a different meaning, and every gemstone has a different meaning. That meaning will obviously be affected by the position of the stone, by the stones which surround it and even by the quantity of a particular type of stone in one selection.

You are an individual and you have to find your own way of giving a reading from crystals and stones. What I have done in the next chapter is to show three basic methods which I have found to be most effective. I would suggest that you practise these until you are familiar with them. After that it is up to you whether you continue to use these methods, whether you wish to adapt them to your own personality in some way, or whether you come upon an entirely different method of your own.

8.
Methods Of Divination

Method One

Sit at a table with your questioner facing you. On the table place your tray of stones and in front of the questioner a black mat. Ask the questioner to select nine stones from the tray, taking as much or as little time as he wishes and selecting those stones which he personally finds attractive. That attraction should be visual rather than by attempting to feel the vibrations of the stones as they may well have been handled by many others before him.

People vary greatly in the way in which they will select the stones. One person will reach in quickly and decisively and will take only a few moments to choose the nine he finds most attractive. Another will move the stones about in their tray over and over again before very deliberately selecting just one stone and then going on to repeat the process for the whole nine. Whichever is the case, you must pay attention to what is going on during this selection as well as observe which stones are handled and considered before being rejected.

It is also necessary to observe the order in which the different stones are selected. The first ones chosen will give you some information as to what is uppermost in the mind of the questioner at that particular time. Suppose he reaches first of all for a lovely mauve amethyst. As you know, this is the spiritual stone and would indicate that the questioner is concerned in some way with his spirituality. If the next two to be chosen are the rose quartz and the aquamarine, you will know that he is also interested in healing, but that he likes to have a logical approach to any subject. Perhaps he would like to use his healing abilities in a practical way; it could well be that he has the makings of a therapist of some sort — something which would allow him to help others to feel better but which would involve a practical and down-to-earth learning process and technique. However, if after selecting the amethyst the questioner went on to pick up the agate geode, you would know that his interest in his spiritual development was more likely to be along psychic or clairvoyant lines.

Sometimes a person will have no difficulty in choosing the first five or six stones but will then have problems when it comes to selecting the last three. This is significant too, for it means that there is only one area of uncertainty and that he feels able to cope with all other aspects of his life.

Watch how the person before you places the stones upon the mat. There are those who will toss them down almost carelessly with no thought for their arrangement. You will probably find that this is a person who has difficulty in putting his own life in order. He may in effect be abdicating responsibility and subconsciously asking you to make some order out of his chaos. Once you have silently noted this fact, it is quite a good thing for both of you if you ask the questioner to take a few moments to sort out the mess of stones upon the mat and to place them in some sort of order so that you may read them more easily.

Another person will select his nine crystals and stones and then take what seems like for ever to arrange them on the mat to his satisfaction. He will make patterns and designs with them and change his mind frequently before finally arriving at a layout which pleases him. This is often a timid and apprehensive person, lacking in confidence and with very little faith in himself.

You can see, therefore, that the period during which the questioner selects his crystals and stones and arranges them on the mat is certainly not one of idleness for the reader. You must watch, note and understand — and you will find that, by the time the nine gemstones are in their final position, you will already know quite a lot about the person sitting before you.

Take note too of the crystals and stones which the questioner rejects. If a stone is picked up and examined before being discarded, it often relates to a situation which is in the process of passing out of the questioner's life. Perhaps there has been a specific problem which he has now managed to overcome; perhaps there is an aspect of his nature which — either because of effort or circumstance — has changed recently. If a particular stone is selected and then at the last moment discarded so that another may take its place, that would indicate that there is an air of change around the questioner at the present time. Perhaps the dark green jasper is chosen and then set aside in favour of the clear red jasper. This would tell you that his emotional life is beginning to change for the better, even though there may have been considerable heartache in the past. You then have to go on to see which stones surround that red jasper on the mat in order to understand the circumstances of the change.

Crystals and stones seem to be of particular value when it comes to dealing with feelings and emotions. Most other forms of divination can give information about facts and events — as of course the crystals can too — but there seems to be no other method which is as indicative of thoughts and feelings of either the questioner or anyone with whom he is directly concerned.

Although the person who comes to consult you will have done so in the hope that you will be able to give him some guidance and information with regard to the future, I always feel that it is beneficial to tell him a little about what is around him at present and what has gone before. If you are correct on these topics you will reassure the questioner that you are not inventing a story to please him but that you are actually receiving information from the crystals and the stones. You will not be able to offer instant proof of what you say with regard to events which are yet to take place, but if you have been able to describe what is and has been around him, he is far more likely to believe what you tell him about the future.

Now is the time to turn your attention to the nine crystals and stones as they lie before you on the mat. Is there a predominance of any particular stone there? This would indicate that the questioner is concerned about one topic above all others. Suppose you found three or four pieces of tiger's eye in front of you. You would know that independence was of great importance to the person having the reading. Look at the stones which surround those pieces of tiger's eye. Do you see the turritella agate representing a change of employment? Perhaps he is thinking of setting up in business for himself. Do you see the multi-coloured variegated jasper with its blotches of red, green and yellow? In that case it could be that the independence arises because of the ending of an unhappy relationship. Remember that it is not just the meaning of each individual stone which is significant but also its position on the mat and its relationship to various other stones around it.

Once you have looked at what seems to you to be the most important part of the reading, it is time to turn your attention to the remainder of the stones on the mat. Study each one in turn and decide whether it seems to stand alone or whether it is linked to any of the stones which surround it.

Suppose there is one area where you feel unsure in your own mind about the meaning of the stones before you. All you need to do is ask the questioner to concentrate upon the stone or stones which baffle you and then to select three further stones from the ones which remain in the tray. These three extra stones should be able to supply you with the additional information you require.

Let us take an example. Suppose the questioner had placed on the mat in splendid isolation an amethystine agate which, as you know, represents a move of home. You may feel that there is more significance to this stone than just that bald statement of its basic meaning. If he selects three further stones and they happen to be a turritella agate, a labradorite and a moss agate, you would know that the move is likely to arise because of a change of job, that it will probably take him overseas and that, despite any misgivings he may have, he will be quite content with the final outcome.

It is also possible, of course, that the anxiety felt by the questioner will not be for himself but for another person about whom he cares and for whom he feels concern. If he is closely attached to someone, whether it is a member of his family or a close friend, he could well have picked up an anxiety that they were feeling. If that is so, then it should become clearer when he selects the three extra stones.

Once you have dealt with the crystals and stones which have been placed upon the mat, it is always a good thing to ask the person before you whether he has any specific question in mind which he would like you to answer. If he has, then you must remove all the crystals from the mat and put them to one side — but not back in the tray — and ask him to choose five further stones from those which remain in the tray. I usually ask the questioner to tell me in very general terms which aspect of his life it is which is causing him concern. I do not want him to give me too much information — apart from the fact that it makes reading objectively far more difficult, it also makes the questioner doubt the validity of the reading if he thinks that he has given you too solid a foundation upon which to build. It is quite sufficient for this purpose for him to say 'my working life' or 'my emotional life' or even 'my youngest son'. The rest will be made apparent when one turns to the five stones he has selected.

In a moment I shall give you a sample case history of a reading I have given (changing the name of the questioner) so that you may more clearly understand how that reading progresses. Before that, however, here is a checklist for you to refer to when you do your own readings:

Step-by-step guide to Method One

1. Ask the questioner to select nine stones from those in the tray. Take note of those he handles and rejects. Also of the method of selection and the order in which the stones are chosen.
2. Study the arrangements of the stones on the mat. If necessary, ask the questioner to arrange them in some sort of order.
3. Look at the largest group of stones and notice whether there is a predominance of any type of stone. Work on these first.
4. If there is any stone which stands alone or of whose significance you are unsure, ask the questioner to focus his attention on that particular stone and then to select three further stones from those in the tray. Study these for your additional information.
5. Remember to give some information about the past and the present in order to give the questioner confidence in your ability and in the validity of readings from the stones in general.
6. Ask the questioner whether he has in mind a particular question. If so, ask him to give you a *general* idea of the topic and then to select five further stones from those in the tray.
7. Always do your best to send the questioner away with hope and a feeling of faith in the future. Even if there are problems ahead, there is bound to be a solution eventually and it is your task to help him towards this solution.

Case history using Method One: Marilyn

When Marilyn first came to see me she was a young woman of about 29. She sat at the table looking at me nervously from wide blue eyes set in a pale and anxious face. As we spoke she continuously twisted a handkerchief in her thin hands. She explained to me that she had never before had a reading of any sort and that she did not know what to expect but that she felt she had to have some sort of help and guidance. I asked her to look at the stones in the tray and to select the nine which most appealed to her and then to place them on the mat.

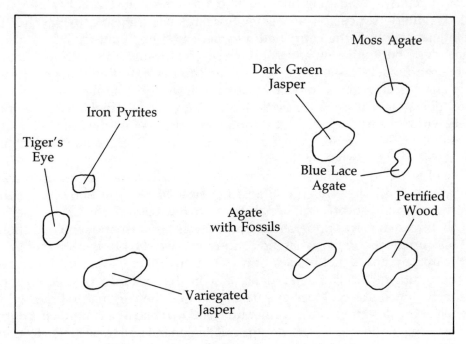

Figure 16: Marilyn (1)

The first stones to be chosen by Marilyn were those which form the group on the right of the diagram. Without hesitation she reached for the dark green jasper of anxiety; this was followed in fairly quick succession by the agate with fossils and the piece of petrified wood. Next came the blue lace agate which Marilyn turned over in her hand and stroked gently with her finger before placing it on the mat. Then she paused. Running her fingers through the stones in the tray she lifted out a very large piece of variegated jasper, placing it on the left-hand side of the mat. Close by that jasper she put the golden iron pyrites and a large piece of tiger's eye. Then she stopped and looked at me. 'You have only chosen eight stones,' I said to her. 'Please could you pick one more.' Marilyn selected a piece of moss agate, held it for a few moments and then replaced it in the tray. She hesitated and then picked up that same piece of moss agate and placed it, as you will see from the diagram, near the group containing the dark green jasper.

I looked at the stones in the order in which they had been chosen. The dark green jasper, which indicates anxiety and concern, was placed beside the agate with fossils referring to money, and the piece of petrified wood, which symbolizes the law or lawyers. I felt that Marilyn's present anxious state had been brought about because of financial difficulties and that possibly legal help would be needed to resolve them. Also in that particular group of stones was a blue lace agate, indicating concern about a young girl — possibly her daughter. When I told this to Marilyn she explained that she had recently been divorced after her husband had left her for another woman and that she was having great difficulty in managing financially to look after herself and her four-year-old daughter, particularly as her ex-husband was refusing to pay her the maintenance which the court had awarded. She had contemplated seeking legal advice but until now had not done so, preferring to live in hope that her ex-husband would come to his senses and 'do the right thing' for his former wife and his little girl. I told her that I felt from the reading that legal help would be sought in order to settle the situation and — looking at the moss agate which she had selected after some hesitation — that the outcome would bring her peace of mind, even though I realized she might have her doubts about this at that particular time.

Then I turned my attention to those stones shown on the left of the diagram. The variegated jasper was the largest I possessed. As you know, this stone always indicates severe emotional hurt on the part of the questioner. In Marilyn's case this was understandable as it was her husband who had chosen to break up the marriage. What did concern me was the presence in that group of the iron pyrites, denoting lack of trust, and the tiger's eye, which can refer to the positive side of independence but in this instance seemed to mean desperate loneliness on Marilyn's part. I felt that the very natural hurt which she felt at the desertion of her former husband had had a profound effect upon her, causing her to lose her trust in people in general and to become alone and somewhat introverted. When I put this to her, Marilyn told me almost angrily that she would never put herself in a position again where she could be betrayed and that she would prefer to live the rest of her life alone and without emotional involvement. Then she looked at me and her eyes filled with tears. She told me that she did not really mean what she had said, that she was desperately lonely and wanted to think that at some point in her life she would find love again — but did not know whether she would ever be able to trust another man as she had had no idea of her husband's deception until he had told her of his long-standing affair and his desire for a divorce.

I removed the stones which were lying on the mat and set them aside. Then I asked Marilyn to select a further five stones from those in the tray and to place them on the mat.

Let us look now at the second set of stones chosen by Marilyn. First came the serpentina and the turritella agate together. These were placed, as you will see from the diagram, almost in the centre of the mat — and beside them she put a tiny piece of red jasper. After thinking for several moments and running her fingers through the stones still in the tray, Marilyn chose an amethystine agate and a larger piece of red jasper, and she placed these two in the corner of the mat, well away from the other stones she had selected.

The serpentina can mean more than one thing, as we have already discussed. In this case however, I felt that it referred to the summertime. The turritella agate placed beside it told me that it was likely that Marilyn would be

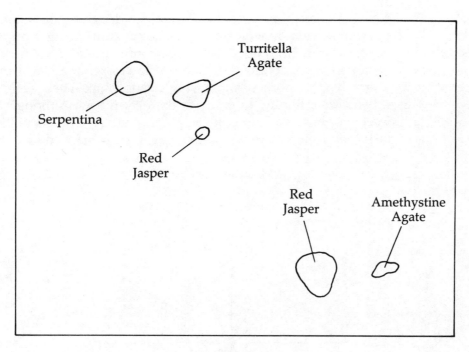

Figure 17: Marilyn (2)

either getting or changing a job the following summer. Marilyn told me that she was not working at the present time (which was late November) as her daughter attended a playgroup in the mornings and had to be collected at midday — and there did not seem to be any jobs available which would fit in with this routine. The little girl would be five in a few months time, however, and would be going to school in the spring. This would give her mother the opportunity to look for work of some sort (probably secretarial which was what she was used to) as her days would then be free. The very small piece of red jasper placed alongside those other two stones told me that there was a possibility that Marilyn could well become involved by the summer with someone whom she would meet through this new job. 'Oh, I don't think so,' was her reaction. 'I don't think that could happen.'

Next I looked at the two stones you will see in the bottom right-hand corner of the diagram. The amethystine agate was the smallest in my collection, so I felt that the move was not likely to take place for about 18-24 months but, looking at that large, unblemished red jasper, I told Marilyn that I thought the reason for that move would be a happy loving relationship.

I had two letters from Marilyn at intervals after she had been to see me. The first arrived at the beginning of the following year and told me that she had, indeed, sought legal advice with regard to her financial difficulties and that her ex-husband had received a reprimand from the court and his maintenance payments to her and her little girl were now arriving regularly.

About a year later I received yet another letter from Marilyn. She told me that she had found a secretarial post soon after her daughter had started school. She had become friendly with one of the accountants in the firm and he had slowly helped her to build up both her self-confidence and her trust in men again. Over a period of months they had become closer and closer until now they were involved in a deep and loving relationship. They had just become engaged and were hoping to marry in the near future. Although the move I had foreseen for her had not yet taken place, they were in fact looking for a house which they could share when they began their married life.

Method Two

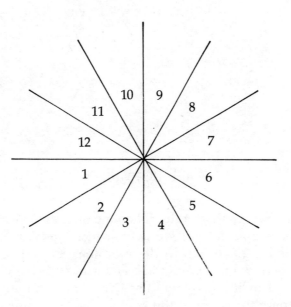

Figure 18

On the table between you and the questioner place the chart illustrated in Figure 18. It can be drawn on card or embroidered on dark fabric. Ask the questioner to select nine stones and to place them in the marked sections on the chart. He may put each of the nine stones in a different section or two or more of them in any section he may choose.

As with Method One, it is essential to pay attention to what is happening during that time when the selection is being made. Is the questioner definite in his own mind about what he wants. Is he hesitant and unsure of himself? Perhaps he chooses the first four or five stones without any difficulty whatsoever, but finds it harder to select the remainder. All these points will help you to understand the questioner and the problems which concern him. So too will taking note of any stone the questioner may reject after having handled it and considered it for some time.

Each section of the chart refers to a different aspect of the questioner's life and that fact, in conjunction with the actual crystals and stones selected, will give you sufficient information with which to offer help and guidance to any who may need it. What follows is a list of the sections, as they are numbered on the chart, and their meaning within a reading:

1. This section relates to the questioner himself. It tells you something about his personality and how he is likely to be affected by future events. Suppose he were to place a tiger's eye and a dendritic agate in section 1: This would tell you that he will be satisfied with the outcome of a particular situation and that his feelings of independence and the ability to stand on his own two feet would be increased.
2. Section 2 gives you some information about the questioner's future financial position, particularly when this is an area about which he has been concerned. It also deals with the acquisition of material possessions. If he has selected a piece of iron pyrites and placed it in section 2, you will know that he will have to take care to ensure that he is not deceived or led astray by someone else in such a way that he loses financially. If, however, you find a piece of turritella agate in that section, you will see that a change of job is likely to bring him financial improvement.
3. This is one of those sections which has more than one possible meaning. It can refer to short-distance travel, so that, if an amethystine agate is placed there, it is likely that the questioner will be moving home, but it does not seem that he will be moving to a place which is very far away from his existing home. It can also refer to communication and education — perhaps the presence of a piece of petrified wood will tell you that there are to be some dealings with a lawyer. In addition, section 3 concerns the immediate family — brothers, sisters, cousins, etc.
4. This is the section which deals with the home environment. It can refer to the actual house the questioner lives in or to the domestic situation which exists there. It is also the section which relates to the questioner's mother so that, should he select a dark green jasper and a bloodstone and place them in section 4, you would know that he was feeling some concern for his mother's health.
5. This section too has more than one area of relevance. It can refer to the questioner's creativity, to children and young people, to romance, or to speculation. You may wonder how you are supposed to know which meaning is appropriate in any particular reading but, provided you have worked on the earlier section in the book which dealt with the development of the intuition, you will instinctively know which aspect of the questioner's life you are dealing with. Any reading you may give is a combination of two things: the meanings of the actual stones selected and their position on the chart or mat, and also the psychic ability or intuition of the reader. If you have not developed your intuition before you begin to give readings to others, the help you will be able to give will be far less, and inaccuracies are more likely to occur.
6. Here we are concerned with the questioner's work and/or career.

Naturally, if you find that he has placed the lovely citrine quartz in this section, you will be able to tell him that his career is likely to go in a new direction. If he selects a glowing red jasper and sets it down in section 6, there is a possibility that he will become emotionally involved with someone with whom he will first have a working relationship.

Another meaning of section 6 is health. This can be the health of the questioner himself or the health of someone about whom he cares. It is up to you to take note of the other stones chosen and their position in the reading so that you can give the best guidance possible.

7. Here we are concerned with close relationships of all sorts, whether they are emotional or business. If the word 'relationship' is used, many people immediately think of love affairs, but of course one has many close relationships in a lifetime — with family, close friends, business partners, etc. Section 7 can concern any or all of these. For example, should the questioner place a pretty little blue lace agate in section 7 alongside a dark green jasper and a turquoise, you would know that there have been problems concerning his relationship with a young woman — possibly his daughter — but that the future would hold peace of mind and tranquillity in this area of his life.

8. This section refers to the area of sex, birth, and death, as well as to questions of life after death. It is in this section that you are most likely to find such crystals as amethysts, agate geodes and rose quartz as these all have to do with spiritual awareness in one form or another. Section 8 is also concerned with inherited money, big business, and with crime.

9. Long-distance travel is a feature of section 9, as is study and higher education. If the questioner selects a turritella agate and a labradorite and places them within the confines of section 9, the probability is that he will be changing his job for one which takes him overseas.

10. This section deals with the questioner's environment and his social standing. It also refers to his father as well as to situations concerning his career or profession.

11. Section 11 concerns groups and organizations and the part they play in the questioner's life. For example, if you were to find in this section an amethyst and a rose quartz, it could well be that the questioner was going to join some sort of healing group.

12. This section deals with escapism in its positive or negative form. If you find in it a piece of rutilated quartz or a white and amber agate, you will know that the questioner is creative and artistic and his form of escapism brings beneficial results. If, however, you see a tektite or several dark green jaspers, it is likely that his form of escapism relates to depression and melancholia.

When the questioner has chosen his crystals and stones and placed them on the chart, look at them carefully. Are they in groups or do you find them isolated and seemingly unlinked? It is easiest to deal first with those which have been arranged in groups, either within the confines of the same section or within the area of two oppositive sections (i.e. 1 and 7, 2 and 8, 3 and 9, etc.), as you will frequently find that these are linked in meaning. Perhaps there will be one (or more) red jasper in section 7, indicating new and happy romantic involvement. Look opposite to section 1 to see the effect this relationship will have on the personality of the questioner. If there is a dark green jasper, perhaps he is anxious or afraid to make close commitments; if you see instead a moss green agate, you will know that the relationship will bring him peace of mind.

If you are confused by a single stone set in isolation on the chart, ask the questioner to concentrate on that particular stone and then to select three further stones and to place them anywhere on the chart he chooses. You will find that the addition of these three extra stones make things much clearer to you.

Ask the questioner whether he has a specific question he would like to ask you. If he has, he should then select five stones and place them on the chart. (All the original stones should be removed before this is done in order to avoid confusion — but do not put those stones back in the tray: the questioner must make his selection from those remaining).

I normally try to ensure that the questioner has put to me *all* the questions he has in his mind before he leaves me, although I will only deal with them one at a time. Many people are so anxious about the outcome of a particular situation that it is only after they have developed confidence in you and your ability they will actually put into words the one thing which really worries them.

Once again, to make things easier for you when you come to practise giving readings for yourself, here is a step-by-step checklist for Method Two.

Step-by-step guide to Method Two

1. Ask the questioner to select nine stones from those in the tray, taking note of the ones he handles and rejects as well as those he finally chooses.
2. Once he has chosen his nine stones ask him to place them on the chart, either in nine separate sections or with two or more in any one section. None should straddle the line between two sections; if any do happen to be placed on a line rather than in one of the spaces, you must ask the questioner to decide in which section he wishes to lay the stone.
3. Study the stones on the chart. Deal first with the largest group of stones.
4. If any stone stands alone, or if you are uncertain as to the relevance of a particular stone, ask the questioner to concentrate upon it and then to select three further stones from the tray, placing them wherever he wishes on the chart.
5. When you have dealt with all the stones and their positions on the chart, ask the questioner whether he has any specific queries. If he has, deal with them one at a time to avoid confusion. In each case ask the questioner to give you a *general* idea of the area of concern and then to select five stones

and place them on the chart. Make sure that all the original stones have been cleared from the chart before this is done. Set these stones to one side — do not return them to the tray.
6. Remember that it is your duty to make sure that the questioner is able to go away after the reading with hope in his heart for the future. You have a great responsibility and should never send someone away feeling that there is nothing to look forward to. This does not mean that you may not point out possible problems but that you should try and look on the positive rather than the negative aspects of any situation.

Case history using Method Two: Edward

Edward was a 62-year-old widower with silvery-white hair and a delightfully dry sense of humour. He had never had a reading but he came to see me because his daughter had been impressed by the accuracy of a reading I had done for her some time previously.

Edward took his time over his selection of the first nine stones. Carefully and methodically he turned over one stone and then another, looking at them closely, admiring some and rejecting others. Eventually he had chosen his nine stones and I asked him then to arrange them on the chart in any way he wished.

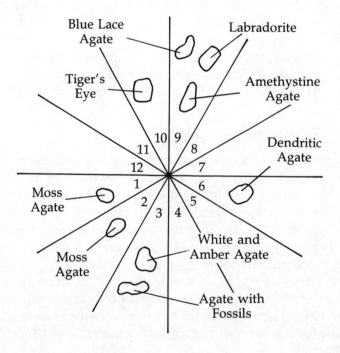

Figure 19: Edward (1)

As you will see from Figure 19, Edward placed three stones in section 9 of the chart — the section which concerns long-distance travel. The stones were the amethystine agate, denoting a move of home, the labradorite, indicating that the travel was likely to be overseas, and the blue lace agate, referring to a young woman. I told Edward that I thought it likely that he would be moving overseas in the near future and that there was a young woman concerned in the move. He confirmed that his only daughter now lived in the United States and, as he had recently taken early retirement, he had been contemplating moving to California to be near her. He had not finally made up his mind but he was interested to note that it had arisen in the course of the reading.

Next I turned my attention to section 3 — which refers to communication. There he had placed a piece of white agate, surrounded by an amber border, and an agate with fossils. Looking at these I suggested to Edward that perhaps he was considering taking up writing and that it appeared that he would be able to earn some money by doing so. As opposite sections are often linked, I felt that any writing project was more likely to be undertaken after he had made the move than before.

Although the other stones in the reading were each placed in a separate section, I felt that these were all linked with the main question in Edward's mind, so I did not feel the need to ask him to select further stones at this point.

Let us look at those stones one at a time. Edward had selected two moss agates and placed one in section 1 and one in section 2. This told me that from the point of view of financial reward, and also from the point of view of personal fulfilment and satisfaction, Edward would find contentment in the future. This contentment appeared to relate both to the move overseas to be near his daughter and also to taking up writing as a full-time career. In section 6 Edward had placed the glowing and positive dendritic agate, indicating that both his health and his work in the immediate future would be fine. Finally, in section 10 there was a large piece of tiger's eye, referring to that independence which Edward was seeking in his personal and professional life. Although he was anxious to live nearer to his daughter, he did not wish to become a burden to her in any way — and the position of that tiger's eye enabled me to reassure him that he would maintain his independence.

I asked Edward whether he had any questions and he told me that I had, in fact, answered the only query which had been in his mind. Would I mind, however, just putting his mind at rest on one other point? I cleared the stones from the chart and asked him to choose a further five stones and place them wherever he wished on the chart.

As you will see from Figure 20, Edward's area of concern seemed to be around section 6 where he had placed a tektite, a bloodstone and a turquoise. Now section 6 can refer to work and career — but I felt that we had already covered this topic. It can also refer to health and, particularly as the bloodstone was present, I thought that this was where Edward's anxiety lay. Indeed, although he had tried to minimize the anxiety he felt, the presence of the piece of tektite showed me that there was more than a small worry in this area. However the third stone he had chosen was the turquoise — and this is always a good sign.

The Crystal Workbook

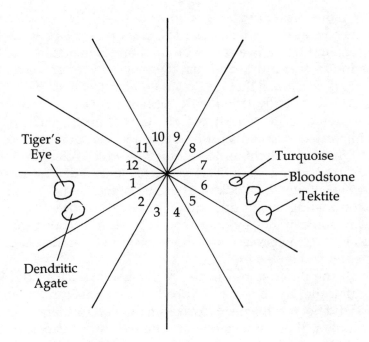

Figure 20: Edward (2)

The tektite, as you will have seen from the list, refers to a deep anxiety — in some cases it could even mean depression. The bloodstone alongside the tektite was yet another indication that Edward's concern was for his health but the pretty little turquoise showed tranquillity in this regard, so I did not feel that there was any particular cause for concern.

In section 1 Edward had placed the dendritic agate and the tiger's eye, showing that his worry to do with his future health was more from the personal point of view than the financial — even though, if he was contemplating living in the United States, the latter must also be a consideration. Section 1, as you will have read, concerns the personality of the questioner and I felt that he was more concerned about losing his independence and having to lean too heavily upon his daughter than he was about any medical bill he might be called upon to face. Edward confirmed that this was, in fact, so. He did not want to travel half-way across the world to be near his daughter only to become a burden on her at some future date. I was able to reassure him that, because of the presence of the positive dendritic agate and the independent tiger's eye, it did not look as though this would be the case.

Method Three

Sit at a table with the chart between you and the questioner. Make sure that the chart is placed with the box number 1 furthest away from the questioner

Methods of Divination

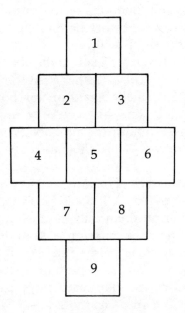

Figure 21

and the box number 9 closest to him. Now ask him to select nine stones from the tray as usual, but in this case he must choose each one with a particular box number in mind. For example, he must actually think to himself 'Which stone or crystal would I like to place in box number 1?' and then must choose and position that stone before going on to number 2 — and so on.

It is interesting to note the different ways in which people approach this task. Some will simply concentrate on the boxes in order and place one stone in each, starting with 1 and ending with 9. Others will choose stones for the different boxes in random order. But of course you know that this is not really a random choice. Watch closely and see which are the first box numbers to have stones placed in them; this will give you some information about the problems which concern the questioner at that particular time. For example, you will learn that box number 8 concerns areas of difficulty or restriction in the questioner's life. Suppose he selects a bloodstone first and places it in box number 8 — you will know that he is concerned about his health and how that will affect his future.

Below are the general meanings attributed to each particular box on the chart, together with some examples of what would be indicated when a particular stone or crystal is placed within their boundaries:

1. This box concerns the questioner himself and in particular the direction in which he is likely to be concentrating his energies in the near future. Perhaps he will select from the tray the delicate rose quartz and place it in box number 1. This would tell you that he has an interest in healing which is likely to be developed in the months to come.

2. This section refers to the emotions of the questioner — his hopes, fears, anxieties, etc. It also reveals future happenings concerning his immediate family or his close friends. Box number 2 would also show you possible areas of change — but these are likely to be changes brought about by the actions of the questioner himself rather than things which happen to him. If he were to place a turritella agate in this box, for example, it would indicate a change of job, but this would be unlikely to be due to being dismissed or made redundant; the probability is that the questioner will decide to change his job and will deliberately set out to do something about it.
3. Box number 3 refers to areas of expansion in the questioner's life. This can be in any of a number of forms. It could be a practical expansion in that he may find himself better off financially — or even with an increased family! It could be a form of mental expansion — as in the case of someone who is going to begin a course of study, for example. It could even be an expansion of spiritual awareness — perhaps in the case of someone who is going to sit for spiritual development. Suppose the questioner were to place in box number 3 a piece of moss agate. You would know that his peace of mind was going to grow. You would then have to turn your attention to the other stones and their positions on the chart in order to find the reason for this — whether there was likely to be a change in circumstances which would bring it about or whether the questioner would simply be learning to cope more easily with any problems in his day-to-day life.
4. Sudden change brought about by circumstances over which the questioner has little or no control — that could be the meaning of box number 4. It can also refer to independence, either in its happy connection as when he is able to set up in business for himself, or less happily when he feels himself to be alone in the world. Whichever is the case, box number 4 is a very important section to consider, particularly if it happens to be the one on which the questioner concentrates first of all. Perhaps he feels drawn to choose an agate with fossils and place it in box number 4 — as you know this stone represents an increase in money. It is possible that a sudden financial boost will enable him to do something he has previously been unable to do.
5. This section concerns the intellect, studying, writing and communication. It is often of particular importance when a person who has come to consult you is a student, as it is within this section that you will find the information which will enable you to give him some sort of guidance as to his future career — although of course it is never your task to *tell* him what he must do. It is up to each individual to make his own decisions in life; all you can do is point out the possibilities and probabilities which lie ahead. If the questioner happens to choose a piece of petrified wood to place in box number 5, for example, it could be that he is considering studying law.
6. There are two main areas to be dealt with by box number 6. The first is the love life of the questioner, and the second is his creativity. If he selects a

citrine quartz and places it within this box, it could be that he is about to enter a new phase in his romantic life — possibly with someone he has not yet met. If, however, he chooses a fluorite octahedron and puts it in box number 6, that could well be a sign that he expects so much of himself on a creative level and is so dissatisfied with his achievements in this area that he prefers to do nothing at all.

7. Box number 7 concerns the psychic and intuitive side of the questioner and his spiritual life. It also indicates areas of sensitivity. If, for example, he chooses the white and amber agate, referring to writing, and places it in this section, it could be that he is either capable of spiritual writing or that he is likely to write about intuitive subjects.

8. This section will indicate in which areas of the questioner's life he is most likely to encounter difficulties and restrictions. Suppose he selects from the stones and crystals in the tray the cubic form of the piece of iron pyrites; this would show you that it is his tendency to gullibility or to placing his trust in the wrong people which is going to cause him problems.

9. This box and the stone or crystal placed within it will give you guidance as to the area of his life with which the questioner feels dissatisfied and where he would like to make changes. To discover whether those changes are in fact imminent, you will have to take into account any other stones placed on the chart. For example, if you see an agate quartz in box number 9, it may be that the relationship between the questioner and his son is not all that it could be. A moss agate or a dendritic agate placed in box number 2 would enable you to inform him that the situation is likely to take a turn for the better.

Once the questioner has selected his nine stones and placed them in the various squares marked out on the chart, you will find, however, that most of the sections are open to more than one interpretation and this is where your own psychic ability comes to the fore. With a little practice, and provided you have already worked through the techniques for psychic development set out earlier in this book, you should instinctively feel certain in your own mind which area you are talking about. Should you be unsure of a particular interpretation, do not hesitate to ask the questioner to concentrate, say on box number 4, and to select another stone and place it beside the one already there in order to give you more information to work upon.

When you have dealt with the matters which arise from the initial selection of stones, you must, as always, ask the questioner whether he has any specific query. The method in this case is slightly different to the other two you have already learned. First of all, as usual, you must remove all the crystals and stones which were originally placed on the chart. Then, if the questioner tells you his problem concerns his marriage, ask him to concentrate upon box number 6 and to select three further stones and to place them within the confines of that particular box. You should then be in a position to give him the guidance and assistance he requires.

Step-by-step guide to Method Three

1. Place the chart on the table between you, making sure that box number 1 is furthest away from the questioner and box number 9 is closest.
2. Ask the questioner to select 9 stones, choosing one for each particular box — although not necessarily in ascending order. Take note of the order of selection as this will show you which areas of his life concern him the most.
3. If you feel uncertain as to the precise aspect of life referred to in any of the boxes, ask the questioner to select one additional stone and to place it beside the one already in that particular box.
4. Before answering any specific query the questioner may have, remember to clear the existing crystals and stones from the chart and to set them to one side.
5. Ask the questioner to give you the general heading under which his query arises. Then ask him to concentrate on the particular box relating to this area (e.g. box number 7 if the problem concerns his psychic development) and to select three further stones or crystals from the tray.
6. Remember, as with methods 1 and 2, your task is to give the questioner information which is as honest and as accurate as possible while still making sure that he is left with feelings of optimism and hope for the future.

Case history using Method Three: Sylvie

When Sylvie first came to see me she was 31 years old — a dark vivacious

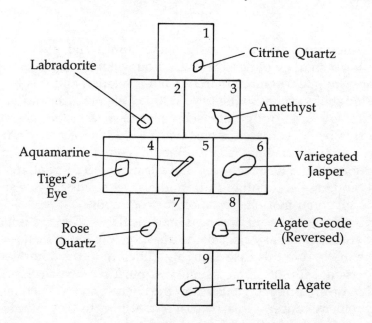

Figure 22: Sylvie (1)

Methods of Divination

Frenchwoman who had come to England some eight years earlier as the bride of a London businessman. Her English was perfect and she had no trouble at all in expressing herself or in understanding what was said to her. I decided to use Method 3 for her reading so I duly placed the chart in front of her and asked her to select nine stones, one for each of the boxes. I told her that she could choose the stones in any order she wished but that each one must be picked with a particular box in mind.

Running her long slim fingers through the stones as they lay gleaming in the tray, Sylvie immediately chose a citrine quartz which she placed in box number 1. This indicated, of course, that there was about to be a change in her life, but until I saw the other stones chosen I would have no idea about the area of her life where that change was to take place.

The next crystal to be chosen was a piece of rose quartz which Sylvie placed in box number 7. The rose quartz is the healing stone and box number 7 refers to the questioner's psychic or spiritual awareness, so I felt that Sylvie certainly seemed to have healing ability.

After this, in fairly rapid succession, Sylvie selected an amethyst which she placed in box number 3, a turritella agate for box number 9 and a tiger's eye for box number 4. These gemstones and the boxes in which they were placed led me to tell her that I felt she had a desire to change her job for one which would give her more independence — possibly even the opportunity to work for herself. As I was speaking Sylvie nodded and, almost unconsciously, picked up a long cool piece of aquamarine which she placed in box number 5. The combination of this choice with the selection of the amethyst in box number 3 told me that, although she had the ability to work as a spiritual healer, Sylvie's nature would lead her to need a practical down-to-earth outlet for her healing abilities. You would find this combination of stones in the reading of anyone who works in a practical way to make others feel better — from a nurse to a Samaritan. I explained all this to Sylvie and she said that, yes, she did wish to change her job and to work for herself. She wanted to help people in some practical way and was considering taking a course to become a reflexologist. However, having divorced her English husband the previous year, she knew that she had to be able to earn her living as well as satisfy the need to be of help and she was anxious to know how I felt the future would turn out if she did in fact leave her fairly well-paid job in the commercial world and become a therapist.

I asked Sylvie to select one more stone for box number 1 and another for box number 7. The first one she chose was a dendritic agate and this she placed in box number 1, indicating that she was likely to be very pleased with the positive outcome of the path she was about to take. In box number 7, beside the rose quartz which had been one of her first choices, Sylvie placed an agate with fossils, which inferred that she would be able to make her living out of her healing ability and that, in fact, her financial state would improve in the future.

Next I turned my attention to those of the original nine stones which I had not yet considered. In box number 2 was a lovely piece of labradorite telling me of a change to living overseas. I was surprised at this initially until I realized that for Sylvie, England *was* overseas. She told me that after the divorce she had temporarily considered returning to live in her native France but that she had now changed her mind and decided to remain in England where she had made her life for the past eight years.

In box number 8 — the section dealing with difficulties and restrictions — Sylvie had placed an agate geode. As you know, this is the stone which tells of intuitive abilities and psychic development. I was however a little concerned that she had put it in that particular box and also that she had placed it face downwards so that its glittering mauve interior was hidden from view. I felt that Sylvie in some way doubted her abilities as a healer and that this very doubt could in some way restrict her progress. When I put this to her she explained to me that her upbringing had been one which concentrated solely on the logical approach to life and that her family and teachers had had no belief in spiritual and intuitive powers and certainly had not felt that they had any part to play in the way one earned one's living. Although one part of her knew that she did, in fact, have psychic and healing abilities, Sylvie found it very difficult to go against this background and wanted to be sure that she was not fooling herself. I explained to her that the selection of the agate geode, together with the amethyst and the rose quartz, was a sure sign that these abilities were in fact present in her and that that combination, together with the practical course which she intended to take, should make her an excellent therapist of any sort.

When Sylvie had chosen her original stones, she had had little difficulty with the first eight but box number 6 had obviously caused her some concern. At first she felt that she did not wish to put anything in that box but when I gently insisted she chose a very large piece of variegated jasper — glowing with red, green and gold. Of course, this stone indicates severe emotional hurt which, although in the past, is still having some effect on the questioner. She had placed it in box number 6, referring to love and deep emotions, so I felt that the one area of her life where Sylvie really felt vulnerable was in her emotional or love life. I knew that she had been divorced but felt that it must have happened in such a way as to make her feel that she had lost her confidence in this aspect of life. This feeling was reinforced when I asked her to choose another stone and place it beside the variegated jasper. After a few moments' thought she selected a piece of iron pyrites which she also placed in box number 6. Iron pyrites, of course, indicates a lack of trust and a degree of gullibility. I felt that Sylvie had lost the ability to trust either herself or others in emotional matters and even that there was a sense of anger at her own stupidity. When I explained this to her, Sylvie told me that the divorce had been set in motion after she had discovered that her husband had been having an affair with her own 'best friend' for several months. She was annoyed with herself because she had no idea that this was happening — indeed she was convinced she had a happy and successful marriage — until she found some letters which the woman had written to her husband. As well as being angry

Methods of Divination

with herself, the break-up of her marriage had made Sylvie determined to remain independent and had convinced her that she would never enter into a deep emotional relationship again.

Having finished the initial part of the reading, I cleared all the crystals and stones from the chart, setting them to one side. I then asked Sylvie whether she had any specific question she would like to ask me and she said that she had. When I requested a general heading for this question, she told me that it concerned her mother who still lived in France. I told Sylvie to concentrate on box number 2 and to choose three stones from those which remained in the tray and to place them in that section of the chart.

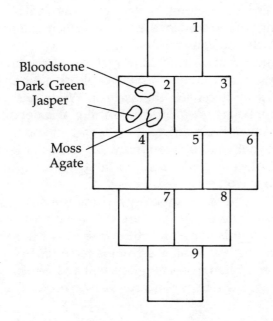

Figure 23: Sylvie (2)

The stones Sylvie chose were (in order of selection) a bloodstone, a dark green jasper, and a moss agate. The bloodstone told me that we were dealing with a health problem. The dark green jasper was an indication of Sylvie's concern and the moss agate showed me that the final outcome would be satisfactory. I told Sylvie, however, that I felt intuitively that this satisfactory outcome was not in the immediate future and that her mother would, in fact, have health problems for several months to come, probably resulting in treatment, whether traditional orthodox medicine or some alternative therapy. Sylvie found this quite amusing as her mother was a strong disbeliever in all forms of complementary medicine so that any treatment she might have was bound to be of the more orthodox variety. She was happy to know, however, that the problem would have a happy solution as her mother's less than perfect health had been one of the reasons why she had contemplated returning to live in France.

Some eight or nine months later, Sylvie wrote to me to tell me that she was half-way through an aromatherapy and reflexology course which she was thoroughly enjoying. Her mother's health had not been at all good and the old lady had finally had to have quite a serious operation but, after some days of anxiety, she had begun to make a strong recovery and was now resting at home before coming to visit Sylvie for a holiday later in the year.

You now have three entirely different methods of giving a reading using crystals and stones. Try all of them on your friends and family and see which of them you find most suitable for your abilities and your personality. It may be that you will need to adapt one or more of the methods in a way which appeals to you. This is not wrong; we are all individuals and, provided you keep to the meanings of the stones, the way in which you choose to use them must be one to which you can readily relate.

I would not suggest that you use any method of reading which relies on intuitive interpretation for yourself. Some people do attempt to do so but they usually only end up by deceiving themselves. When one gives a reading to someone about whom one knows little or nothing, it is not difficult to say what one really feels. It is an entirely different matter when you try to give yourself a reading. You *know* what you hope for and what you want for the future — and that knowledge makes it very easy to fool yourself into believing the interpretation to be as you wish it to be.

Remember too that you are not merely reciting the meanings of the crystals and stones as they lie before you. You are calling on your own psychic ability to help and guide you. It is important, therefore, that before you begin to give readings of any sort you first of all practise the techniques for developing your own intuitive powers. Then and only then will you be able to use your own crystals and stones as divinatory tools to enable you to give help to others.

9.

Dowsing

Dowsing is an art which has been around since the days of our ancestors. In former times men would dowse in order to find sources of water. They used a forked stick which would normally be made of hazel as this was considered the most appropriate wood for the task. The dowser would hold the stick and walk slowly over the ground. When he came to a place where water was present, however far below the surface, the stick would begin to move, to shake — or even in some cases to break — in his hands.

No one has yet been able to give a universally acceptable explanation for dowsing and why it works. None the less there is little doubt, even among scientists, that it *does* work. In fact, a great deal of money has been saved by certain oil companies who now use dowsers to guide them to the best place to drill. Dowsing can involve the use of a pendulum, a rod, or even a twig, and it can be used to answer any question you may have. As well as searching for water or oil, you can use a pendulum to help you to find a lost item, to check upon health, and to test for allergies or vitamin deficiencies. And the beauty of dowsing is that it is something you can so easily try for yourself — often with amazing results.

The easiest form of dowsing is with a pendulum, and for this purpose the quartz crystal is ideally suitable. Dowsing is the link with the subconscious mind and, as has already been shown, quartz crystal can intensify the link between the conscious and subconscious mind, enabling you to attune with your own higher levels of consciousness.

The first thing, naturally, is to acquire your own piece of quartz crystal to use as a pendulum. This should only be a small piece of crystal, perhaps two inches in length. The crystal you use should be selected by you personally using any of the methods set out earlier in the book for choosing crystals intuitively. The next step is to cleanse and charge your crystal in the usual way, and then to mount it on a chain, preferably one made of silver as the combination of crystal and silver has the correct vibrations for dowsing. Once your crystal has been charged, it should be kept for this purpose alone and not used for anything else. Neither should it be handled by others as it is your own personal link with your higher self. Should it be handled by anyone else you must recleanse and recharge it before using it as a pendulum again.

A pendulum can only give the answers 'yes' or 'no' to any question put to it and the first thing you have to do is discover which way it swings to give you these answers. My own pendulum swings clockwise for 'yes' and anticlockwise for 'no' (although there are occasions when this is reversed, but we will deal with this at a later stage) — but it may not be the same for you. Below is a method for discovering the meaning of each rotation of the pendulum for you:

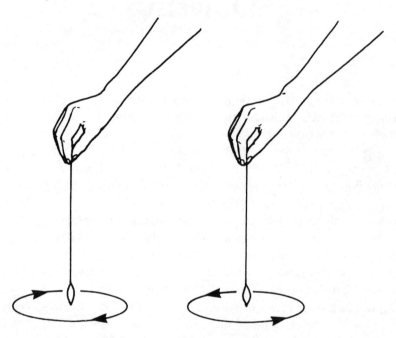

Figure 24: Testing for direction of 'Yes' and 'No'

1. Hold the end of the chain between the thumb and forefinger of one hand. Extend the arm, as in Figure 24, keeping it as relaxed as possible. Make sure that the arm slopes slightly downwards as this causes less muscular tension. Remember that the more relaxed you are, the better the results will be.
2. Swing the pendulum gently backwards and forwards in a north/south direction.
3. Ask a question to which you know the answer. For example, I might ask 'Is my name Ursula?' and take note of the swing of the pendulum. After a few moments it should begin to make ever-increasing circles in either a clockwise or an anticlockwise direction and you will know that that is the direction which for you indicates 'yes'.
4. To confirm your findings, steady the pendulum, and swing it again gently in the north/south direction. Now ask it a question to which you know the answer is 'no'. If you have done it correctly, the crystal should begin to swing in the opposite direction.

Remember:
- Never swing the pendulum deliberately in a circular formation or use any force.
- You do not have to ask the questions aloud; you merely need to think them.
- The process may take some practice, particularly if you are a fairly tense person as the tension of your own muscles will affect the swing of the pendulum. Some people find that the pendulum is also capable of swinging from side to side (i.e. east/west) and this usually indicates that there is no positive answer to the question being posed and that the questioner is in fact on the wrong track entirely. It is as well to take a few moments to reaffirm the direction of the swing of the pendulum on each occasion you wish to use it as it can vary. Although it will normally remain the same, it has been found that if one is particularly tired or upset, the direction can be reversed.

Once you have worked out the direction of the swing of your own pendulum, you can use it to answer any question which would receive the answer 'yes' or 'no'. Obviously you cannot ask the pendulum 'Do I need more or less calcium?' as there is no positive or negative answer to that particular question. However, correctly used, the ways in which a pendulum can be employed are many and various.

Analytical Dowsing

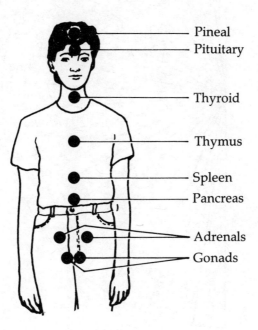

Figure 25

The Crystal Workbook

When you think about it, a great deal of medical diagnosis, whether on humans or animals, involves a certain amount of educated guesswork on the part of the practitioner. An animal cannot tell you what is wrong with it and human beings are often vague and may well confuse more than one set of symptoms. A pendulum can be of great assistance here, although it should be used as an aid to conventional diagnosis rather than a replacement for it. In addition, a pendulum can often diagnose a problem before the symptoms occur.

Analytical dowsing requires a suitably charged crystal pendulum, a large copy of the diagram illustrated below and a second charged quartz crystal which is compatible with that used as the pendulum.

Figure 25 indicates the glandular centres of the physical body and, by use of the pendulum, you should be able to tell whether or not those glands are functioning efficiently. The dowser should set his pendulum to swing north/south and then, touching each of the centres on the diagram in turn with the point of the second crystal, he should silently ask the question 'Is there any

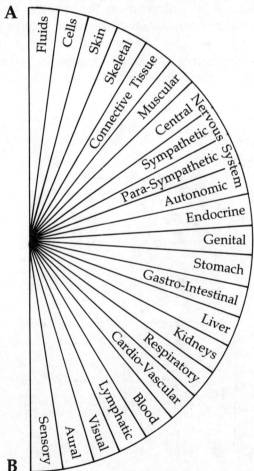

Figure 26

malfunction of this particular gland?', taking note of the answer given in each case.

A second method of analytical dowsing is to employ a similar technique using Figure 26. There are two ways of using this particular diagram which, as you will see, contains the names of areas of the physical body which may be malfunctioning.

As before, swing the pendulum, gently in a north/south direction. While doing this, touch the first section — the one marked 'Fluids' — with the point of the second crystal and ask the pendulum whether this is an area which requires help or treatment. Once the answer to that question has been noted, proceed to the second question, and so on.

Alternatively, swing the pendulum in a north/south direction over the line AB and mentally ask for an indication of any area in the patient which may be in need of treatment. The pendulum will, after some moments, change direction to swing along the line indicating in which area to look.

Vitamin Deficiency

The Western diet, consisting as it does of so many highly processed and fatty foods, does not always supply the correct vitamins in the necessary quantities, but it is extremely difficult for the layman to assess how much of any particular vitamin he is in fact receiving from his normal diet. Using the list which follows, and holding the pendulum in one hand while touching each vitamin and mineral in the list with the point of your second crystal, it is possible to ask whether your daily food intake supplies all that you need of each item in turn. Having done this, you will know in what way you should supplement your diet with additional vitamins and minerals. (The list also gives you details of the function of each particular vitamin and mineral).

Vitamin	*Function*
Vitamin A (carotene)	Essential for normal growth, for the structure and functioning of the skin and for the health of the eyes.
Vitamin B1 (thiamine)	Important for growth and for the correct functioning of the heart, muscles and nerves. Also affects the metabolism.
Vitamin B2 (riboflavine)	Necessary for the sense of general well-being. Also for health of eyes, mouth and hair.
Vitamin B3	Growth of all tissue and health of hair and skin.
Vitamin B6 (pyridoxine)	Protein metabolism. Correct functioning of muscles and nerves. Health of skin.

Vitamin B12	Protein metabolism. Health of skin and of nerve tissue.
Biotin	Health of skin and proper functioning of nerves and muscles.
Choline	Essential for proper functioning of the liver. Prevents build-up of fatty acids in the body.
Folic acid	Essential for red blood cell formation and for growth and division of cells.
PABA (para-aminobenzoic acid)	Helps red blood cell formation and assists bacteria in forming folic acid.
Inositol	Necessary for correct functioning of the liver. Prevents build-up of fatty acids.
Niacin	Health of skin. Function of stomach, intestines and nerves.
Vitamin C	Growth cell activity. Health of teeth and gums.
Vitamin D	Formation of bones and teeth. Regulating calcium and phosphorus metabolism.
Vitamin E	Function of nerves and muscles. Normal reproduction.
Vitamin F	Blood coagulation. Important for normal glandular activity.
Vitamin K	Essential for clotting of blood.

Trace Elements	*Importance*
Cobalt	Maintains red blood cells.
Copper	Formation of red blood cells.
Iodine	Regulates body metabolism. Promotes correct functioning of the thyroid gland. Prevents goitre.
Iron	Haemoglobin production. Promotes growth.
Zinc	Insulin synthesis. Necessary for male reproductive fluid. Aids the healing process.

Minerals	*Importance*
Calcium	Essential for strong bones and teeth. Helps

	functions of the heart, muscles and nerves. Aids clotting of blood.
Chlorine	Regulates acid/alkaline balance of the body. Aids digestion.
Magnesium	Helps the body to use fats, carbohydrates, proteins and other nutrients.
Phosphorus	With calcium — helps to strengthen bones and teeth.
Potassium	Aids correct functioning of kidneys, nerves and heart muscles.
Sodium	Essential for correct functioning of muscles, nerves, blood and the lymph system.
Sulphur	Acts in formation of body tissues.

Allergies

Dowsing to discover the nature of the substance to which you may be allergic is a rather more difficult and often vague process. The list of things to which it is possible to become allergic is endless, so there is no guarantee that you will be able to arrive at the correct conclusion. Having said that, however, there are many hundreds of documented cases where a sufferer has in fact been able to discover the cause of his or her allergy by means of a pendulum.

It is probably advisable to carry out this process in two separate stages. In the first place it is necessary to discover whether your allergic reaction is brought about because of something which you take internally, something with which you come into physical contact, or something which is in the atmosphere. This at least helps to reduce the number of possibilities when it comes to making a more detailed examination.

So, to start with, you need a somewhat larger version of the very simple diagram illustrated below. You can either draw this on a sheet of paper, or some people actually prefer to have three separate small boxes which they can lay on a table or other flat surface.

Substances taken internally	**Substances in physical contact**	**Substances in the atmosphere**

Hold your pendulum and set it to swing in the north/south neutral direction over each box in turn. As it does so, ask whether your allergies are caused by the substances indicated in that particular box. It is advisable to repeat the process with all three boxes even if you receive an affirmative answer from the box number one as it is of course possible to be allergic to more than one thing.

Having decided which group or groups of substances you are dealing with, the next stage is to try and narrow it down somewhat. To do this you will need to make lists of substances which could come under those three headings and test each one separately with your pendulum to see whether you get a positive or negative result. The lists given below are intended merely to start you off on your quest — naturally they will vary from person to person. They do contain, however, the most common causes of allergy.

Substances taken internally
- Milk
- Eggs
- Sugar
- Wheat
- Salt
- Caffeine
- General dairy products
- Red meat
- White meat
- Fish
- Alcohol
- Medication

Substances contacted physically
- Dust
- Grass
- Animal fur
- Skin cream/make-up
- Man-made fibres
- Wool

Substances in the atmosphere
- Smoke
- Dust
- Pollen

Testing for Blockage of the Chakras

As we have already seen, in order for the mental, physical and spiritual self to function at maximum efficiency, it is necessary to have all the chakras clear and unblocked. One of the main difficulties, however, particularly for someone who is comparatively new to chakra work, is that it is not easy to determine where in the body the blockage lies. Even if you feel sure that there is one particular chakra (or possibly more than one) which is not clear and functioning as it should, because they are so closely inter-linked it can be confusing when one tries to find the source of the blockage. This is one area where the pendulum can be of invaluable help — particularly when the pendulum is made of quartz crystal. Your subconscious mind will know where the chakra blockage is even though your conscious mind may not. And since the movement of the pendulum is in fact guided by your subconscious mind and the quartz crystal enhances your own natural thought processes, you should be given very clear information as to where the chakra blockage lies.

Using Figure 27, holding your pendulum in one hand and touching the different chakra points with your second crystal one at a time, all you have to do is ask whether each chakra is clear or blocked and you should receive a definite answer. Once you have your answer you can employ the method described in Chapter 3 to clear the chakra so that the whole system functions efficiently and well.

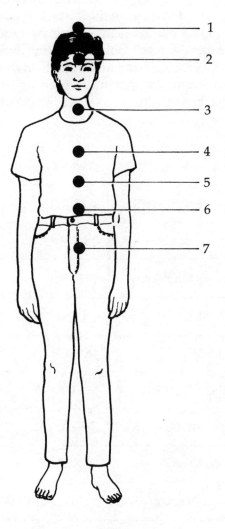

Figure 27: Chakra points

Choosing the Most Appropriate Gemstone for Healing

In Chapter 2, the section on gemstones and crystals and their use in healing, a list was given indicating which gemstone was appropriate for each disease or each part of the body. In many cases, however, the subject may be aware that all is not well without feeling that he has problems in any one particular area. In such cases it can often be helpful to use a pendulum to select the stone which will prove most efficacious in healing that person. Of course it should be remembered that gemstone or crystal healing is intended to *augment* any treatment which may be being received either from complementary practitioners or from orthodox medicine — not to replace it.

Divide a tray or a large sheet of card into boxes as illustrated in Figure 28, and in each section place one of the healing crystals described in the list starting on page 29. If you do not have the various types of crystal mentioned, it would be possible to achieve the same effect by writing the name of each clearly in the various boxes. Use your pendulum over each box in turn to reveal to you which crystal would be most suitable in the existing case. (Do not forget, of course, that it may be that more than one type of crystal will be needed, so try all the boxes even if you have already received a positive indication.)

AGATE	AMAZONITE	AMBER	AMETHYST
AQUAMARINE	AVENTURINE	AZURITE	BERYL
BLOODSTONE	CARNELIAN	CAT'S EYE	CHALCEDONY
CHRYSOCOLLA	CHRYSOPRASE	CITRINE	CORAL
DIAMOND	EMERALD	GARNET	HEMATITE
HAWK'S EYE	JADE	JASPER	JET
LAPIS LAZULI	LOADSTONE	MAGNETITE	MALACHITE
MOONSTONE	OBSIDIAN	ONYX	OPAL
PEARL	PERIDOT	PYRITES	QUARTZ
RED CORAL	RHODOCROSITE	ROSE QUARTZ	RUBY
RUTILATED QZ.	SAPPHIRE	SERPENTINA	SMOKEY QZ.
SODALITE	SPINEL RUBY	STAUROLITE	TIGER'S EYE
TOPAZ	TOURMALINE	TURQUOISE	ZIRCON

Figure 28

Finding Lost Items

Suppose you have mislaid a piece of jewellery somewhere around the house and you do not know where it can be. Because on a subconscious level we never forget anything we do, your pendulum should be able to help you find it. Begin by listing the rooms in the house and testing them one at a time with your pendulum to find out where to begin your search. Once you have received information as to which room your item is in, draw a diagram of the room containing all the major pieces of furniture — similar to the one in Figure 29. If you then close your eyes and hold your pendulum over the diagram and you will find that it begins to swing in one particular direction indicating one specific area of the room. Now all you have to do is search in the area indicated by the pendulum and you should find your property with little difficulty.

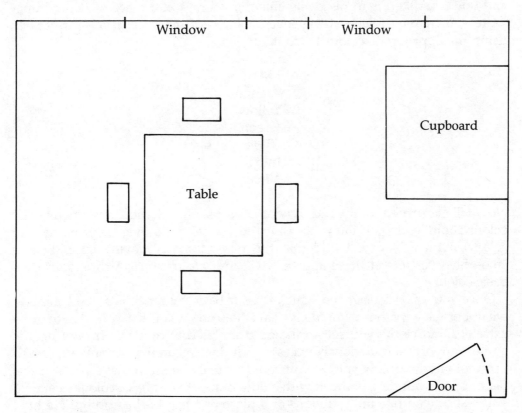

Figure 29

There are those who use a similar method to try and find missing people or items with which they have had no physical contact. Although there have been a few reports of this technique being successful, the number is small indeed and it is possible that those 'successes' are really due more to the psychic or intuitive ability of the dowser than to the pendulum itself. While there is a logic to finding an item which you personally have misplaced because the details are stored somewhere within your own internal memory bank, you will have no hidden memory in your subconscious mind to guide you when it comes to looking for a missing person or hidden treasure.

Colour Therapy

Colour therapy, as defined by Max Luscher, Rudolf Steiner, Isdore Friedman and others, is often thought of as a new innovation whereas in fact it has been recognized, albeit in limited circles, for thousands of years.

Colour is an energy with a measurable frequency and wavelength and it can strongly affect human beings in many different ways. It can bring feelings of peace and serenity; it can stimulate mental activity or physical energy; it can cause definite and recognizable emotional problems. But — and perhaps this is the most important point of all — colour can heal. Each of us has a basic colour and your first step is to discover which is yours. On a piece of white paper write your name in each of the seven colours of the spectrum — preferably using the appropriate colour to do it.

Red
Orange
Yellow
Green
Blue
Indigo
Violet

Now take your pendulum and, having asked silently to be shown your basic colour, hold your pendulum so that the crystal rests over each name and colour in turn. You will find that the pendulum will swing strongly and smoothly over one of these and that will indicate to you which is your own basic colour.

Once you have determined which is your basic colour, it is a good thing to wear it or have it around you daily. This does not mean that you have to go to extremes and clothe yourself in indigo from head to foot! A sense of degree and of proportion is obviously necessary. It does mean that, should you feel a sense of fatigue or low spirits, you could actually begin to make yourself feel better by surrounding yourself with a little more of your basic colour — even if you just concentrate upon ribbons or a piece of fabric of the appropriate hue.

From time to time it is possible to become deficient in one or more colours. For, although each of us has one basic colour, our auras are composed of all the colours in the spectrum in varying proportions. If we become deficient in one or more of the colours we can find ourselves feeling moody, depressed or even unwell — feeling 'off colour' in fact. If you feel that this could apply to you at any time, use the chart in Figure 30 to indicate where the deficiency lies. Swing the pendulum gently along the line AB and ask it to show you which colours you lack. You will find that the pendulum will change direction so that it swings along the line of a specific colour. Make a mark alongside that colour and then ask the pendulum to show you if there are any other deficiencies. Repeat this process until the pendulum continues to swing along line AB continuously. It is not unusual to find that you are indeed deficient in two or even three colours; although unless there has been some violent trauma in your life, it is unlikely to be more than three.

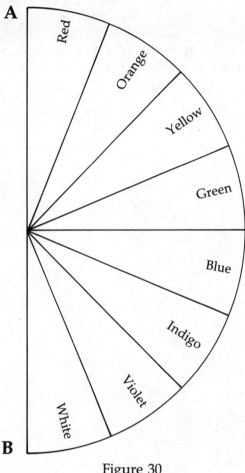

Figure 30

There are various ways of counteracting these deficiencies, some of which are indicated below. For more information you could consult such books as *Colour Healing* by Mary Anderson or *The Luscher Colour Test* by Dr Max Luscher.

- Place a lamp of the appropriate colour in a light fitting and sit under it for a period each day;
- Drink colour-charged water — place a colour filter over a glass of water and leave it in the sun for at least two hours before drinking it.
- Visualize the appropriate colour or colours flooding into your body for a regular period of time each day until you notice some improvement.
- Concentrate upon a piece of cloth or ribbon of the colour in which you are deficient.

- Focus your attention upon a stone or crystal of the appropriate colour and set aside a period each day for meditation upon this crystal.

Whichever method you choose, the process should be repeated daily for about three weeks. At the end of that time, use your pendulum once more to test for colour deficiency. You should find that all is now well. If by any chance there is still a need for a little more absorption of a particular colour, just continue the process until your pendulum indicates that the colours of your aura are evenly balanced.

10.

Gem Elixirs

As this is a subject of which I have no personal knowledge or experience, I am most grateful to Phyllis and David Lovell of Crystal World for their assistance, both with information and help they have given me and also for allowing me to reproduce certain parts of their literature. All lists in this chapter are reproduced with their kind permission.

The use of gem elixirs as an aid to physical, emotional and spiritual healing originates from the United States of America. The pioneer of work in this field is Gurudas, and he in turn received the greater part of his knowledge and information through Kevin Ryerson, a trance channel of many years experience. This information was given through the trance channel in much the same way that Edgar Cayce used to receive his knowledge. Although he has been working on and striving to perfect the gem elixirs, Gurudas is the first to admit that there is more yet to be learned. Indeed, in his book *Gem Elixirs and Vibrational Healing, Vol. I*, he states:

> 'The information presented in this book is not meant to be the final word. It is my sincere wish that this book stimulate greater awareness, understanding and clinical research of these remedies. In the coming years, much new information on gemstones will be revealed by various sources. This is especially true with quartz crystal technologies.'

There are some 144 gem elixirs presently available. These can be purchased or you can follow the method set out in this chapter and make them for yourself. In some cases it is recommended that one should take drops of one single elixir, whereas in others it is the combination of several different elixirs which produces the desired result. A comprehensive list of elixirs available is set out at the end of this chapter, together with the conditions they are said to help. As with any other form of healing, however, it is not being suggested here that you should substitute the taking of gem elixirs for any other treatment you may be receiving. The intention of the elixir is to enhance and accelerate the healing process.

Preparing Gem Elixirs

Gem elixirs are prepared in a similar way to natural flower essences. You will require:

- A clear glass bowl. This should not have any patterns or designs on it, the colour of which could affect the elixir during the course of its preparation.
- Glass bottles.
- Glass funnels.
- Labels.
- Distilled water taken from a natural spring. If using bottled spring water, ensure that it is the non-carbonated variety.
- Selected gemstones. These must be uncut, unpolished and ideally crystalline in structure.

All glass items should be new and should not be used for any other purpose. They should also be sterilized by placing them in hot water (for example, in an enamel or glass — but never aluminium — bowl).

Gurudas also advises that certain very powerful gemstones should be placed under the sun or the moon for a length of time of up to two weeks.

There are two acknowledged methods for the preparation of the gem elixirs — the Sun method and the Boiling method. Either can be successful and you should merely select the method which either appeals to you the most or which fits in most easily with your particular routine.

The Sun Method

1. Place the chosen mineral in natural sunshine in order to activate its inherent properties. At the same time it is necessary for you to sit in silent tranquillity and meditation.
2. Place the gemstone in the centre of a bowl of distilled water and surround it with quartz crystals or rubies.
3. Place the bowl on a natural surface, such as wood or grass. Leave the gemstone in the water for two hours. It is best to use this method on a cloudless day and preferably in the morning.

    If you decide to cover the bowl, do so with a piece of glass on which there are no colours or designs. It is possible to leave the bowl uncovered but you will then have to remove any object which may have blown onto the surface of the water (this can be done either with clean hands or with a quartz crystal).
4. Pour the liquid into a bottle and label it. Gurudas recommends that, if you are using new containers, they should be placed under a copper pyramid for between 30 minutes and two hours.
5. If intending to reuse the bowls and equipment, it is important that you sterilize them between each process. You should also be sure to wash your hands each time in order to avoid inadvertent mixing of the elixirs.

The Boiling Method

Cleanse the gemstone in sea salt or powdered quartz. Then boil it for ten or fifteen minutes in a clean glass bowl. Ideally this should be done at sunrise, at noon, or even (with certain gemstones) with the setting of the sun. What you will produce is a 'mother essence' — the elixir will be produced by placing seven drops of the mother essence in a bottle of pure water.

In his book, Gurudas also sets out details of the cleansing, protecting and storing of gemstones and gem elixirs. He does not suggest that they are a universal cure but that they should be used in conjunction with either complementary or orthodox medicine.

Commonly-Used Elixirs

The list which follows details the most commonly used of the elixirs and the ways in which they can help those who take them.

Abalone — An elixir for strengthening of the heart; also for treating any spinal degeneration disease.

Agate (Botswana) — This elixir is particularly effective in high-pressure oxygen therapies, used to treat tumours, neurological and skin tissue regeneration and lung damage.

Agate (Carnelian) — This preparation is used to treat anorexia nervosa.

Agate (Fire) — Influences the entire endocrine system and restimulates the memory cells.

Agate (Moss) — This agate can ease lymphomas, Hodgkin's disease, diabetes. It can also be used to ease allergies, kidney disorders and liver problems.

Alexandrite — Has an impact on the nervous system, spleen and pancreas. Central nervous system diseases, leukaemia, lymph and spleen-associated diseases are alleviated. Low self-esteem and difficulty in centring the self are important clues to the need for this elixir.

Amazonite — The activities of the heart and solar plexus chakras are aligned, which also aligns the etheric mental bodies. This elixir has important ethereal properties.

Amber — A strengthening of the thyroid and the inner ear.

	People needing amber may exhibit memory loss, inability to make decisions or eccentric behaviour.
Amethyst	Calms down stressful situations, allows clarity of thought, is good for headaches and migraine. Gives confidence, stimulates those who lack vision, and helps to have a greater attunement with God.
Aquamarine	Reduces fear and anxiety and the inability to express oneself. Excellent for all throat and upper chest conditions.
Aventurine	Alleviates psychosomatic illness, anxiety and buried fears (particularly those which originate from the first seven years of childhood). This elixir is applicable to psychotherapy in that it encourages the development of emotional tranquillity and a more positive attitude to life.
Azurite	This is effective in the treatment of diseases of the bone, such as arthritis and spinal curvature.
Azurite Malachite	This elixir has an impact upon the liver, skin and thymus, as well as upon muscular dystrophy and cirrhosis of the liver.
Beryl	This is applicable to the intestinal tract and the cardiovascular system. It is also used to treat hardening of the arteries. The over-analytical or hypercritical person could benefit from this elixir.
Bloodstone	Assists with the improvement in condition of the bone marrow, spleen, heart, testicles, ovaries, cervix and uterus. This elixir also generates a higher state of spiritual consciousness.
Brass	Eliminates toxins throughout the body; stimulates hair growth; aligns the vertebrae. Helps scalp and skin diseases.
Bronze	Helps produce red corpuscles.
Calcite	Increases mental capacity. Used for those wishing to try astral projection.
Chalcedony	Stimulates bone marrow and increases the production of red corpuscles.

Chrysocolla	Strengthens the lungs, thyroid and coccyx. Eases stress and hypertension and balances emotions. Use this elixir when practising breathing exercises for more control. It also amplifies the throat chakra.
Chrysoprase	This elixir's primary focus is upon the prostate gland, testicles, fallopian tubes and ovaries.
Citrine	Removes toxins from lower area of body.
Copper	Can be used for a wide range of inflammatory conditions, such as arthritis, inner ear and intestinal disorders. It strengthens the pineal and pituitary glands and aligns the five lower chakras.
Coral	It balances the entire personality; use for emotional calm, attunement to nature and to increase expressive abilities and creativity. All meridians and etheric bodies are balanced.
Diamond	This is extremely powerful for removing blockages and negativity which may be interfering with vibrational remedies. It removes blockages in the personality and is used in cases of anxiety and insecurity.
Emerald	Balances the heart chakra, stimulates the meridian points, balances the emotions, strengthens will power and improves the memory.
Fluorite	Strengthens the teeth. Eases bone tissue and dental disease. In a mouthwash this elixir helps to prevent dental decay.
Garnet	The first chakra is stimulated. This elixir is ideal for people who have a tendency to be self-centred.
Gold	Ideal for balancing the heart chakra. As the heart is critical to the circulatory flows of the body, it is the master healer. It aligns higher spiritual thoughts, and increases the ability to give and receive love.
Haematite	Helps the blood cleansing function of the kidneys.
Herkimer Diamond	This elixir releases stress and tension — in

particular the type of stress which leads to malignant tumours. In a similar fashion to the diamond, the Herkimer radiates energy.

Jade	Helps the individual to become more articulate. Gives courage, wisdom, sensitivity and increases psychic abilities.
Jasper (Green)	This elixir is used to promote healing on all levels and aligns intuitive forces.
Lapis Lazuli	Effective in treating tonsillitis. Its impact extends to the larynx and bronchial passages. Releases buried emotions and anxieties. Improves meditation.
Lodestone (Negative and Positive)	Increases the biomagnetic forces in the body. Can be used in magnetic healing. Makes the aura more sensitive and, by strengthening the aura, many holistic therapies are explained. It balances the male and female.
Magnetite	Stimulates the entire endocrine system; chakras are aligned and deeper meditation is stimulated.
Malachite	Assists in correcting irregular menstruation. Fertility increases, stomach ulcers are reduced.
Meteorite	This is a very important elixir and can be taken at any time. It protects against planetary radiation, ley lines and geopathic stress, and it is good for protecting the aura.
Moonstone	Helps to control emotions, particularly those which cause anxiety and stress. It is a female stone and should be used for pelvic disorder.
Obsidian	Balances the stomach area and intestine and muscle tissue in that area.
Onyx	Regenerates the heart, kidney and nerve tissue.
Opal (Dark)	This opal affects the ovaries, testicles and pancreas as well as helping to release depression and acting as a grounding element for the emotional body. With this elixir the thought force is amplified.
Opal (Light)	Helps ease autism, dyslexia, epilepsy and visual problems. Helps the individual who is seeking higher inspiration.

Gem Elixirs

Pearl — Alleviates all emotional imbalances. It is a powerful elixir for treating emotional difficulties. Emotional stress affects the stomach and lower back and can also manifest itself in various stress-related diseases.

Peridot — Helps to gradually remove all toxicity from the body. Gives a more positive emotional outlook on life and increases patience.

Platinum — Decreases arrogance, depression, pride, stress; also memory loss brought about by shock or tragedy.

Pyrite — It is a digestive aid for the abdomen and upper intestines. Aids the production of enzymes.

Quartz (Smokey) — Good for sexual problems in both sexes. Proper release of the kundalini energy takes place with this type of quartz. Meditation with smokey quartz removes unclear thought forms.

Quartz (Rutilated) — Tends to reverse ageing disorders associated with a lower immune system. Stimulates inactive or unused parts of the brain. Increases clairvoyance and develops inspiration to experience the highest spiritual teachings.

Quartz (Clear White) — Amplifies the crystalline properties in the body. Alleviates emotional extremes. Improves assimilation of the amino acids in protein. The emotional and etheric bodies are aligned as are the brow, crown, and solar plexus chakras.

Rhodocrosite — Helps to detoxify the kidneys. There is a general strengthening of self-identity and an ability to function better in life.

Rhodonite — This elixir strengthens the inner ear — particularly bone tissue and the sense of hearing.

Rose Quartz — Helps overcome emotional problems such as anger or tension. Increases self-confidence and negates false pride. Balances the emotions and stimulates the heart chakra.

Ruby — Acts on the first chakra which in turn connects to the heart, activates the kundalini, creates balance in spiritual endeavours and amplifies thought power.

Sardonyx	Aids the lungs, larynx, thyroid and the nervous system.
Silver	Stimulates the nerve tissue. The IQ increases and the speech centres are stimulated. Silver is also used in cases of mental imbalance, such as schizophrenia.
Sodalite	Strengthens the lymphatic system. Helps to attain emotional balance expressly for the purpose of spiritual growth.
Spinel Ruby	A powerful general cleanser. Helps to detoxify the system.
Star Sapphire	Aligns the spinal column and improves communication with the higher side of life. Links the mind, body and spirit to bring attunement.
Tiger's Eye	Works on the adrenal glands and cleanses the bowel and bladder areas.
Topaz	Calms the passions. Improves the appetite and is a source of strength when dealing with life's problems. Stimulates the third chakra and helps to maintain equilibrium of newly stabilized emotions.
Tourmaline (Black)	Assists the first chakra imbalance such as arthritis; adrenal disorders are alleviated.
Tourmaline (Rubellite)	Activates the qualities stored in the second chakra. For example, stimulates creativity and fertility.
Tourmaline (Clear)	The third chakra is activated and problems associated with this chakra are alleviated. (For example, digestion and ulcers.)
Tourmaline (Green)	Opens the chakra, regenerates the heart, thymus and immune system.
Tourmaline (Blue)	Activates the throat chakra. Strengthens the larynx, throat and thyroid.
Tourmaline (Cat's Eye)	This is the elixir for the sixth chakra. It stimulates the endocrine system and awakens personal concepts of God.
Tourmaline (Quartz)	Opens the crown chakra. All subtle bodies and chakras are aligned. There is greater attunement

	to the higher self and an increased spiritual understanding.
Turquoise	This is a master healer. It strengthens the entire anatomy and protects the aura.
Zircon	The forces of the pineal and pituitary gland are merged on a physical level. The chakras associated with these two glands are opened and balanced.

Symptoms

For easy reference the lists which follow (taken from the literature of Crystal World) contain symptoms and the appropriate gem elixirs for the treatment of those symptoms. The list is in three separate sections — physical problems, psychological problems, and psychospiritual problems.

Physical Problems

Acidity	Pyrite.
Acne	Moss Agate, Aventurine, Fire Agate.
Addiction:	
Alcohol	Amethyst, Emerald, Jade.
Tobacco	Gold.
Drugs	Emerald.
Ageing	Gold, Silver, Copper.
AIDS	Amethyst, Alexandrite, Garnet, Platinum, Silver, Carbon, Steel.
Anaemia	All Tourmalines, Gold, Garnet, Magnetite, Molybdenum.
Appetite	Citrine Quartz, Amethyst.
Arthritis	Lapis Lazuli, Dark Opal, Clear Quartz, Ruby.
Asthma	Pearl, Tourmaline (Cat's Eye), Amethyst, Emerald.
Backache	Amethyst.
Bladder Weakness	Lodestone, Malachite, Silver.
Blood Pressure:	
High	Pearl, Emerald.
Low	Ruby, Magnetite.
Irregular	Ruby, Pearl.

Bronchitis	Pearl, Azurite.
Burns	Malachite, Pearl, Ruby, Turquoise.
Calcium Excess	Chrysocolla.
Cataract	Malachite, Quartz (Clear), Turquoise.
Circulatory Problems	Emerald, Ruby, Bloodstone, Light Pearl.
Cold Hands and Feet	Ruby, Magnetite, Haematite.
Colds	Pearl.
Colitis	Emerald.
Colour Blindness	Amethyst.
Constipation	Topaz, Ruby.
Convulsions	Copper, Magnetite.
Cough	Topaz, Azurite.
Cramps	Copper, Malachite, Moonstone.
Croup	Topaz, Diamond.
Cysts	Magnesium, Spinel.
Dandruff	Diamond.
Deafness	Moonstone.
Diabetes	Moss Agate, Lodestone, Malachite, Amethyst, Rhodonite.
Diarrhoea	Beryl, Magnetite.
Digestion	Coral, Onyx.
Dizziness	Copper, Kunzite, Lodestone, Magnetite, Malachite.
Dysentry	Emerald, Copper.
Dyspepsia	Emerald, Diamond.
Earache	Diamond, Fluorite, Platinum, Silver.
Eczema	Sapphire, Coral.
Epilepsy	Copper, Gold, Silver, Meteorite.
Fatigue	Ruby, Pearl, Moonstone, Emerald, Diamond.
Fever	Copper, Coral, Gold, Jasper, Kunzite, Ruby, Silver.

Fissure — Anus	Diamond.
Flatulence	Coral, Emerald, Topaz.
Gall Stones	Chalcedony, Magnetite.
Gall Bladder	Copper.
Gastric Ulcer	Emerald.
Glands, Swollen	Moonstone, Aquamarine.
Goitre	Topaz.
Gout	Light Pearl, Topaz, Turquoise.
Gums	Carnelian, Coral.
Haemorrhages	Pearl, all Agates, Garnet.
Haemorroids	Coral, Diamond.
Hair	Kunzite.
Hay Fever	Pearl, Topaz.
Headaches	Haematite, Kunzite, Lodestone, Malachite, Platinum, Amethyst.
Heart Disease	Emerald, Ruby.
Heartburn	Chrysocolla.
Hepatitis	Coral, Citrine, Quartz.
Herpes	Aventurine, Garnet.
Hyperglycaemia	Moss Agate, Amethyst.
Hysteria	Pink Coral, Sodalite.
Impotence	Magnetite, Ruby, Tourmaline (Cat's Eye).
Indigestion	Coral, Pearl.
Infections	Malachite, Obsidian, Ruby, Silver.
Infertility:	
Women	Malachite, Rose Quartz, Ruby, Lodestone.
Men	Pearl, Amethyst, Bloodstone.
Inflammations	Bloodstone, Pearl, Ruby.
Influenza	Topaz, Moonstone, Jet, Gold, Fluorite, Copper, Beryl.
Jaundice	Copper, Coral, Magnetite.

Kidney Stones	Amethyst, Emerald, Jade.
Laryngitis	Lapis Lazuli, Pyrite.
Leukaemia	Bloodstone, Bronze, Copper, Diamond, Obsidian, Opal, Rutile, Ruby.
Leprosy	Diamond, Sapphire, Tourmaline (Cat's Eye), Onyx, Fluorite.
Liver Infections	Lapis Lazuli, Opal, Clear Quartz, Ruby, Beryl, Peridot, Molybdenum.
Lumbago	Lodestone, Magnetite.
Memory	Emerald.
Menstruation:	
Heavy	Sapphire.
Irregular	Tourmaline Watermelon.
Painful	Lapis Lazuli, Opal, Clear Quartz, Ruby.
Meningitis	Pearl, Sapphire.
Menopause	Herkimer Diamond, Malachite.
Mental Debility	Topaz.
Migraine	Amethyst, Fluorite.
Mouth Infections	Jasper, Kinzite.
Multiple Sclerosis	Gold.
Mumps	Copper.
Nails, Weak	Graphite, Sulphur.
Nausea	Emerald, Onyx, Sapphire.
Nervous Exhaustion	Opal, Pearl, Quartz, Silver.
Nervous Depression	Meteorite, Copper, Gold, Silver.
Neuralgia	Coral, Amethyst, Diamond.
Numbness	Platinum.
Obesity	Malachite, Amethyst.
Palpitation	Ruby, Malachite.
Pancreas Infection	Magnetite, Malachite, Tourmaline (Black).
Paralysis	Platinum, Kunzite.
Parkinson's Disease	Kunzite.

Gem Elixirs

Pharyngitis	Pyrite.
Pneumonia	Light Opal and Pearl, Amethyst.
Pregnancy Nausea	Tourmaline Rubellite, Silver, Amethyst.
Pre-Menstrual Stress	Lapis Lazuli, Opal, Clear Quartz, Ruby.
Prostate	Haematite, Magnesium.
Psoriasis	Onyx, Carbon Steel, Pearl.
Rheumatism	Tourmaline (Black), Sard, Malachite, Lodestone, Gold, Chrysocolla.
Scarlet Fever	Copper.
Sciatica	Copper.
Schizophrenia	Meteorite, Amethyst, Rose Quartz.
Scleroderma	Beryl, Chalcedony, Marble.
Sexual Problems	Lapis Lazuli, Dark Opal, Clear Quartz, Ruby.
Shock	Meteorite, Pearl, Diamond.
Spasms	Copper, Jasper, Kunzite, Amethyst, Silver.
Spermatorrhoea	Light Opal and Pearl.
Sterility	Dark Opal.
Stomach Troubles	Azurite, Light Coral, Opal and Pearl, Rose Quartz.
Syphilis	Malachite, Light Pearl.
Tonsillitis	Lapis Lazuli, Pyrite.
Toxaemia	Positive and Negative Lodestone, Clear Quartz.
Typhoid	Coral.
Varicose Veins	Copper, Coral.
Vertigo	Tourmaline (Blue), Diamond.
Vomiting	Coral, Pearl.
Water Retention	Tourmaline (Cat's Eye).
Weight Problems	Lapis Lazuli, Malachite, Turquoise.

Psychological List

Acceptance of Life	Bronze, Copper, Malachite.
Androgyny	Rhodochrosite.

Altruism	Lodestone, Magnesium, Garnet.
Anger	Beryl, Diamond, Peridot.
Anorexia	Black Tourmaline, Rose Quartz.
Anti-Social	Asphalt, Bog-Peat.
Anxiety	Amethyst, Lapis Lazuli, Diamond, Onyx, Light Pearl.
Apathy	Copper, Onyx, Dark Opal.
Appreciation of Arts	Rose Quartz.
Argumentative	Sulphur.
Arrogance	Platinum.
Autism	Copper, Gold, Silver, Tourmaline (Cat's Eye), Sapphire.
Bedwetting	Magnetite (Positive and Negative), Dark Opal, Black Pearl.
Blocking Spirituality	Diamond, Clear and Rose Quartz.
Body Image Problems	Jamesonite, Lapis Lazuli, Malachite, Turquoise.
Bravado	Carbon Steel.
Buried Emotions	Ruby, Turquoise, Lapis Lazuli, Malachite, Topaz, Diamond, Rose Quartz.
Carelessness	Lapis Lazuli.
Caution	Zircon, Beryl, Chrysolite.
Child Abuse	Jet, Enstatite.
Child Pressures	Rutile Quartz, Serpentine.
Compulsion	Silver, Carbon Steel.
Concentration	Diamond, Onyx, Light Pearl.
Courage	Jade, Amethyst, Sardonyx, Sodalite.
Death Understanding	Opal (Dark).
Decision-Making	Diamond, Positive and Negative Lodestone, Ruby, Star Sapphire, Garnet.
Delusion	Chrysolite, Jet.
Depression	Beryl, Lapis Lazuli, Limestone, Sapphire, Amethyst, Onyx, Light Pearl.

Gem Elixirs

Disappointment	Ruby, Platinum.
Discipline, Lack of	Azurite, Bronze, Carbon Steel, Lapis Lazuli, Limestone.
Discrimination	Fire Agate.
Disorientation	Diamond, Onyx, Light Pearl.
Dreams	Diamond, Clear and Rose Quartz, Emerald, Peridot.
Eccentricity	Topaz, Dark Opal, Diamond, Gold, Amber.
Ego	Rhodochrosite, Gold, Carbon Steel.
Emotional Balance	Malachite, Opal, Pearl, Tiger's Eye, all Tourmalines, Ruby, Turquoise, Lapis Lazuli, Rose Quartz.
Enjoys Misery	Gypsum.
Envy	Diamond, Topaz.
Excitement	Copper, Gold, Jet, Kunzite, Platinum.
Expressive Ability	All Tourmalines.
False Hopes	Pyrite.
Family Problems	Green and Cat's Eye Tourmalines, Sardonyx, Ruby.
Father Image Problems	Jamesonite, Lapis Lazuli, Malachite, Turquoise, Emerald, Ruby.
Fear of Death	Copper.
Fear — Hidden	Coral, Emerald, Dark Opal, Light Pearl, Amethyst, Jade, Lapis Lazuli.
Fear of Relationships	Sodalite.
Fear of Sleeping	Rhodochrosite.
Fear of Spirituality	Chalcedony.
Fear — Unnatural	Amethyst, Jade, Light Opal and Pearl, Beryl, Peridot, Diamond.
Femininity Clarified	Azurite, Pink Coral, Light Opal and Pearl, Rose Quartz.
Flexibility	Chrysoprase, Onyx, Rose Quartz, Ruby.
Frustration	Lapis Lazuli, Dark Opal, Clear Quartz, Ruby.

Function Better	Rhodochrosite.
Goodwill	Chalcedony.
Greed	Chrysoprase.
Grief	Ruby, Sardonyx, Tourmaline (White).
Grudges	Ruby, Turquoise.
Guilt	Chrysoprase, Sodalite.
Hallucinations	Amethyst, Emerald, Jade, Lapis Lazuli, Diamond, Dark Opal, Sapphire.
Harmony	Fire Agate, Emerald, Ruby, Green Tourmaline, Gold.
Hostility	Lapidolite.
Humility	Sardonyx.
Hyperactive	Graphite, Sapphire, Black Tourmaline.
Hysteria	Amethyst, Emerald, Jade.
Idealistic	Rutile Quartz.
Imagination	Chrysoprase, Garnet, Silver, Spinel.
Independence	Aventurine.
Inferiority	Gold.
Inhibitions	Carbon Steel.
Initiative	Lapis Lazuli, Dark Opal, Clear Quartz, Ruby.
Inner Perception	Light Opal and Pearl, Tiger's Eye.
Inner Strength	Topaz.
Insecurity	Diamond.
Insomnia	Emerald, Clear Quartz, Peridot.
Integrate with Society	Amethyst, Malachite, Magnetite.
Introverted	Blue Tourmaline, Lapis Lazuli.
Intuition	Malachite, Dark Pearl, Tiger's Eye.
Irritability	Beryl, Lapis Lazuli, Limestone, Sapphire.
Joy	Sardonyx, Ruby, Opal, Chalcedony, Agate, Aventurine, Alexandrite.
Lazy	Beryl.

Gem Elixirs

Leadership	Ruby.
Lethargy	Light Opal and Pearl, Tiger's Eye, Topaz, Coral, Diamond, Emerald, Lapis Lazuli.
Loving Nature	Ruby, Rutile, Amethyst, Jade.
Manic Depressive	Gold, Jet, Kunzite.
Maturity	Abalone, Chalcedony.
Memory	Amethyst, Platinum, Kunzite, Copper, Amber.
Mental Balance	Red Coral, Emerald, Opal, Light Pearl, Diamond, Ruby.
Mental Clarity	Opal, Pearl, Tiger's Eye, Diamond, Rose Quartz, Peridot, Beryl.
Mental Discipline	Sardonyx, Azurite, Opal, Clear Quartz, Ruby, Coral, Emerald, Pearl.
Mother Image Problems	Azurite, Pink Coral, Light Opal and Pearl, Rose Quartz.
Negativity	All Tourmalines.
Negotiation Skills	Turquoise, Ruby, Clear Quartz.
Nervous Tension	Garnet, Ruby, Star Sapphire.
Nightmares	Beryl, Diamond, Peridot, Emerald, Clear Quartz, Tiger's Eye, Rose Quartz.
Obsession	Emerald, Rose Quartz, Tiger's Eye.
Original Thinking	Aventurine.
Over-Aggressive	Lapis Lazuli, Dark Opal, Clear Quartz, Ruby.
Over-Pedantic	Beryl.
Over-Sensitive	Chalcedony, Gold, Magnesium.
Passions Controlled	Lapis Lazuli, Dark Opal, Clear Quartz, Ruby.
Passive	Amethyst, Jade, Light Opal and Pearl, Lapis Lazuli.
Past Life Talents	Rose Quartz.
Patience	Ruby, Turquoise.
Personality Balanced	Diamond, Clear and Rose Quartz, Amethyst, Lapis Lazuli, Onyx, Turquoise.
Possessiveness	Quartzite.

Practical Nature	Ruby, Turquoise.
Pride	Platinum, Rose Quartz.
Protection	Platinum, Rose Quartz, all Opals, Beryl.
Psychosomatic Ills	Malachite, Dark Pearl, Tiger's Eye, Lapis Lazuli, Turquoise.
Purpose, Clarity of	Diamond, Platinum, Silver, Jelly Opal.
Reclusive	Platinum, Garnet.
Receptivity	Beryl, Garnet.
Relationships — Improving	Ruby, Turquoise, Lapis Lazuli, Dark Opal, Clear Quartz, Mother of Pearl.
Religious Tolerance	Amethyst, Lapis Lazuli.
Resentment	Coral.
Responsibility	Gold.
Rigidity	All Quartzes.
Schizophrenia	Red Coral, Emerald, Dark Opal, Light Pearl, Clear Quartz, Peridot.
Security, Sense of	Moss Agate.
Sedative	Lapis Lazuli, Malachite, Light Pearl, Turquoise, Dark Opal.
Self-Actualization	Bloodstone, Rhodochrosite.
Self-Awareness	Copper, Rhodonite.
Self-Centredness	Garnet.
Self-Confidence	Moss Agate, Copper, Gold, Rose Quartz, Rhodonite, Ruby, Sardonyx.
Self-Destruction	Rose Quartz.
Self-Esteem	Lapis Lazuli, Malachite, Turquoise.
Sensitivity	Diamond, Onyx, Pearl, Lapis Lazuli, Dark Opal, Ruby, Azurite.
Sexual Conflicts	Lapis Lazuli, Dark Opal, Clear Quartz, Ruby, Fire Agate, Coral.
Shock	Diamond, Pearl, Ruby, Botswana Agate, Malachite, Turquoise, Sapphire.
Shy	Lapis Lazuli.

Sleep Better	Emerald, Clear Quartz, Peridot.
Speech Problems	Apatite, Amethyst.
Suicidal	Gold, Citrine Quartz.
Superiority, Sense of	Platinum.
Superstition	Emerald, Rose Quartz, Tiger's Eye.
Uncertainty	Bloodstone, Coral, Diamond, Emerald, Peridot, Lapis Lazuli, Malachite, Light Opal and Pearl.
Unforgiving	Rutile Quartz.
Violent	Rose Quartz, Clear Quartz, Aquamarine.
Virtue	Sardonyx.
Willpower	Pink Tourmaline, Amethyst, Garnet, Carbon Steel.
Worry	Coral, Topaz.

Psychospiritual List

Agnostic	Amethyst.
Astral Projection	Azurite, Calcite, Haematite, Citrine Quartz, Sapphire.
Atheism	Amethyst.
Attract Spiritual Growth	Chrysocolla.
Auric Protection	Meteorite.
Awareness of God	Amethyst, Cat's Eye, Tourmaline.
Breathing for Spirituality	Amethyst, Garnet.
Channelling	Bronze, Sapphire.
Clairaudience	Herkimer Diamond, Fire Opal.
Clairvoyance	Beryl, Diamond, Peridot, Rose Quartz, Tiger's Eye.
Conscience	Gold.
Cosmic Awareness	Meteorite.
Creative Visualization	Aventurine, Gold, Lazurite, Lodestone, Magnetite, Peridot, Sodalite, Tourmaline, Cat's Eye, Galena.

Creativity	Calcite, Coral, Graphite, Moonstone, Dark Opal, Blue Quartz, Turquoise.
Devas, Attunement to	Jade, White Quartz, Ruby.
Devotion	Serpentine.
Divine Love	Emerald, Gold, Jade, Ruby.
Earth's Energy, Attunement to	Moss Agate, Alexandrite, Coral, Garnet, Diamond, Lapis Lazuli, all Quartzes.
Emotional Calm for Spirituality	Sapphire, Galena.
Expression of Spiritual Qualities	Citrine Quartz.
Faith	Pearl, Green Tourmaline.
Higher Self Attunement	Diamond, Clear Quartz, Rose Quartz.
Inner Peace	Chrysocolla, Sapphire.
Overcome Karma	Amethyst.
Kundalini	Gold, Dark Opal, Amethyst, Blue Quartz, Ruby, Sodalite, Tourmaline, Cat's Eye.
Love of God	All Agates, Gold, Ruby.
Mantras Activated	Rhodonite, Copper, Gold, Silver.
Meditation	All Quartzes and Tourmalines, Gold, Silver, Lapis Lazuli.
Psychic Abilities	Beryl, Diamond, Peridot.
Mind, Body Spirit	Aventurine, Sapphire.
Mind-Soul Link	Jade.
Perception of Higher Levels	Fluorite.
Practical Spirituality	Carnelian, Fire Agate.
Prophecy	Chrysolite, Obsidian, Amethyst.
Raise Consciousness	All Agates, Quartzes, Tourmalines and Opals.
Sexual Energy Raised	Lapis Lazuli, Dark Opal, Clear Quartz, Ruby.
Soul Recall	Jasper.
Spiritual Balance	Ruby, Sapphire.
Spiritual Courage	Sard.

Spiritual Discipline	Tourmaline, Cat's Eye.
Spiritual Expression	Diamond, Clear and Rose Quartz.
Spiritual Goals	Lapis Lazuli, Dark Opal, Clear Quartz, Ruby.
Spiritual Illumination	Clear Quartz, Ruby, Sapphire.
Spiritual Aspiration	Diamond.
Spiritual Purity Value	Diamond, Pearl.
Spiritual Self-Confidence	Copper, Fluorite.
Spiritual Self-Esteem	Kunzite.
Spiritual Understanding	Diamond, Onyx, Light Pearl.
Spiritual Yin/Yang Balance	Coral.
Telepathy	Beryl, Diamond, Peridot.
Visions	Lazurite, Clear Quartz.
Visions Distorted	Jasper, Amethyst.
Wisdom	Jade, Pink Tourmaline, Zircon.

11.

How to Grow Your Own Crystals

Have you ever considered trying to grow crystals of your own at home? It is not nearly as difficult as you might think. And, although you will not end up with precisely those natural crystals listed earlier in this book, there is something rather special about being part of the creation of your own crystals. The idea is not to replace natural crystals, which of course have their own particular energies and properties but to allow you, the instigator of the growth, to attune with your 'homemade' crystals in a way that can never be possible with crystals whose formation did not concern you in any way at all.

All the equipment needed for this process can be found around the home or easily obtained. The crystals will be grown in a water solution and at normal room temperature, so no special conditions are needed.

Equipment

A one-litre capacity saucepan
A one-litre capacity glass jar
A shallow glass dish
Water. This should be distilled or natural spring water (the bottled, non-carbonated variety) and not tap water as the water needs to be as pure as it is possible to obtain.
Chemicals. For details of these, consult the list which follows. Remember that several of them are poisonous and should be kept well out of the way of children, pets and food. For this reason, too, it is advisable to keep the pans, jars, etc. used for growing crystals alone and not to risk using them for anything else.

Suitable chemicals for growing crystals at home are:

Potassium alum	This is easy to use and easily obtainable as it is commonly used in making pickles. The crystals grown will be octahedral in form.
Sugar	Sugar, of course, is commonly available in every home. It will form crystals quite easily but is

more likely to develop a chain of crystals along a thread than one single specimen.

Copper Sulphate* The crystals formed from copper sulphate will be brilliant blue.

Potassium Ferriayanide* The crystals produced will be a rich deep red.

Potassium Ferrocyanide* Lemon yellow crystals will be produced by this chemical.

Rochelle Salt This substance will grow colourless oblong-shaped crystals.

* Poisonous if taken internally.

(If you wish to colour any of the otherwise colourless crystals, simply add a drop of the type of food colouring used for the icing on cakes to the solution).

Method

1. Make sure that all dishes, jars, pans, etc. are thoroughly cleaned and — even more importantly — very well dried as there must be no grease or dust to disturb the formation of the crystals.

2. Fill the saucepan with water and heat until the water is just above room temperature — do not boil. Stir in a little of your chosen chemical and keep stirring until it has completely dissolved. Continue adding the chemical, a little at a time, and stirring until the substance has fully dissolved. You will eventually reach a stage where the water cannot absorb any more solids (there should, in fact, be a residue on the bottom of the pan).

3. Leave the pan undisturbed for about 15 minutes in order to allow any solids which may still be going to dissolve to do so. Those which cannot be absorbed will remain on the bottom of the pan.

4. Pour about half an inch of the solution from the pan into the glass dish. Place the dish in a clean dry place — well away from the curiosity of children and pets. Put the rest of the solution into the glass jar, cover and set aside. After a few days you will see crystals beginning to form in the bottom of the glass dish. Leave them to grow until they are about a quarter of an inch in diameter.

5. This stage requires care. Trying not to disturb the crystals which have formed on the base of the glass dish, tip away very carefully the liquid from the dish. Pour about half an inch of clean, fresh, cold water gently over the newly formed crystals in the dish an then drain at once again. Now you have to leave the crystals to dry naturally in the air. Do resist the temptation to handle them at this stage.

The Crystal Workbook

6. Wash and dry your hands thoroughly. From the crystals growing in the bottom of the dish, choose one which seems to you to be the largest and the most perfectly formed (perferably one which does not appear to have any smaller crystals growing from it). Take a piece of ordinary white sewing thread and carefully loop it around the crystal, tying it in a knot. (This is quite difficult to do and you may have to make several attempts.) If the shape of your chosen crystal appears to make it impossible, it may be necessary to select a slightly smaller one which happens to have a more convenient shape.

7. Return to your jar of solution which you had covered and set to one side. It may be that, although you covered the jar, a few crystals have begun to form on the bottom of it. Empty the contents of the jar very carefully back into the clean saucepan, taking great care not to disturb any crystals which might have formed in the jar. Now tip out those small crystals from the bottom of the jar, discard them and wash the jar well in clean, fresh water. Return the solution from the pan to the jar and measure off a sufficient length of sewing thread to ensure that the crystal around which you have looped it will be suspended about two inches from the bottom of the jar. (see Figure 31). Tie the sewing thread around a pencil or a small stick and place this across the opening of the jar so that crystal and thread hang down into the liquid.

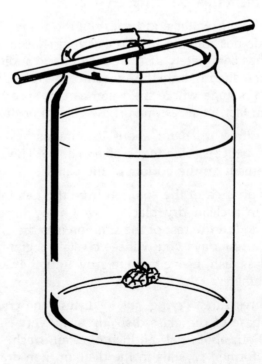

Figure 31: Crystal growing in a glass jar

8. Cover the top of the jar with muslin or gauze or some other material which will keep out dust while still allowing the water in the jar to evaporate. Place the jar in somewhere with an even temperature — perhaps in the airing cupboard. Warmth rather than heat is needed, but the most important factor is the constancy of the temperature. Although it can take several weeks for a crystal of an inch or more in diameter to grow, you should begin to notice some progress within a few days. Although you will want to observe the crystal as it grows, try to disturb the jar as little as possible as this will only cause imperfections in the crystal itself. So, if you can, place the jar in a spot where you can look at it without actually having to move it. A small crystal may be perfect but the larger it becomes the more imperfections you will begin to notice. After all, you are reproducing the processes of nature — and perhaps doing so will help you to appreciate the real beauty of large, perfect crystals.

9. If you wish to grow a crystal larger than about an inch in size you must prepare a second quantity of the solution to replace the original, which will by now have decreased in volume. Do not handle the crystal while changing the solution as the natural oils present in your skin will cause imperfections in the growth of the crystal. Lift the crystal out of the jar by lifting the pencil, hang it inside an empty jar while you empty the growing solution from the jar you have been using and wash and dry it thoroughly. Prepare the new solution exactly as before and, once it has cooled a little, place this new solution in your growing jar. Lift the pencil and replace the crystal in the jar in its new solution and allow it to continue to grow. As the crystal grows larger, its rate of growth will actually become slower.

10. Once the crystal has reached the desired size, remove it from the jar and wash it in *cold* water before drying it gently with a paper tissue. Remove the sewing thread from the crystal. Because these crystals are water soluble, they should be kept in a dry place. Crystals grown in this way are also quite brittle and easily scratched, although they can be handled in the ordinary way.

Perhaps you would like to be able to grow several crystals at the same time. In this case, just make a larger quantity of the original solution and use several different jars, placing a seed crystal in each. It is better to grow each crystal in a separate jar as, should you try to grow several together, the progress will be extremely slow and the results far from perfect.

Of course, one of the advantages of growing your own crystal is that you can begin to charge and programme it from its moment of birth. This could have the effect of making the crystal more effective for whatever purpose you intend to use it — it certainly makes it far more personal to you and compatible with you, your wishes and desires.

Conclusion

You have now been given information on various ways in which you can learn to work with crystals and semiprecious stones. Don't forget that this is only the beginning. No one yet knows the full extent of the scope and power of crystals and their very special energy. You are one of the lucky ones; you are setting out on a great adventure. Use this book as a starting-point so that you become familiar with these precious jewels of nature and aware of what they can achieve. Then, should you wish to progress further, you must use your own intuition to guide you. Who knows — you may be the person who will make the next major discovery about gemstones and what they can do. Whatever happens, you will certainly become more spiritually aware and your higher consciousness will increase enormously the more you come to familiarize yourself with the world of crystals.

Bibliography and Further Reading

Fortune-Telling by Crystals and Semiprecious Stones by Ursula Markham (Aquarian Press) 1987.

Cosmic Crystals by Ra Bonewitz (Aquarian Press) 1983.

Portals of Power by Win Kent and Jesse H. Thompson (International Association of Colour Healers).

Precious Stones, Their Occult Power and Hidden Significance by W. B. Crow (Aquarian Press) 1968.

Crystal Healing by Edmund Harold (Aquarian Press) 1987.

The Healing Power of Colour by Betty Wood (Aquarian Press) 1984.

Crystal Enlightenment by Katherine Raphaell (Aurora Press) 1985.

Clearing Crystal Consciousness by Christa Fay Burka (Brotherhood of Life Inc.) 1986.

Gem Elixirs and Vibrational Healing Vols. I and II by Gurudas (Cassandra Press) 1985/6.

Windows of Light by Randall and Vicki Baer (Harper & Row) 1986.

Suppliers and Organizations

Suppliers of Crystals and Gemstones

Crystal World
Anubis House, George Road, Yorkley, Forest of Dean, Gloucestershire

Craefte Supplies
33 Oldridge Road, London SW12

Natural Earth Crystals
c/o 35, Hargrave Road, London NW5

Falcon Robinson
4, Rushbrook House, Union Road, London SW8 2QY

Spiritual Venturers Association
72, Pasture Road, Goole, North Yorkshire

Sorcerers Apprentice
4/8 Burley Lodge Road, Leeds LS6 1QP

Gem Elixirs

UK
Crystal World
Anubis House, George Road, Yorkley, Forest of Dean, Gloucestershire

USA
Gurudas
PO BOX 2044, Boulder, CO 80306, USA

Organizations

British Society of Dowsers
19, High Street, Eydon, Daventry, Northants

International Association of Colour Healers
33, St Leonards Court, St Leonards Road, East Sheen, London SW14

Institute of Crystal and Gem Therapies
Anubis House, George Road, Yorkley, Forest of Dean, Gloucestershire

Index

fc — illustration inside front cover
bc — illustration inside back cover

acquiring a crystal, 9–10, 74–5
agate, 29, 129
  amethystine, *fc*,79
  blue lace, *bc*,80–81
  botswana, *fc*,81, 129
  dendritic, *fc*,81
  fossil, *bc*,81
  geode, *fc*,77–8
  healing, 29
  moss, *bc*,83–4, 129
  purple, *fc*,84–5
  quartz, *bc*,78
  turritella, *bc*,88
  white and amber, *bc*,89
amethyst, *bc*,9, 29, 78–9, 130
amethystine agate, *see* agate
aquamarine, *fc*,29, 79–80, 130
astrology, 64, 71–2
aura, 21–2

bloodstone, *bc*,30, 80, 130
blue lace agate, *see* agate
botswana agate, *see* agate

chakras, 36–48, 61, 62,120–21
charging, 13–14
citrine quartz, *see* quartz
cleansing, 11–13
colour therapy, 124–6

dendritic agate, *see* agate
diagnosis, 19–21, 117–20
divination (lithomancy), 73–4, 89–90
  methods, 91–4, 98–102, 104–8, 112

dowsing, 113–26

elixirs, 127–9, 129–35

fluorite octahedron, 82, 131

green jasper, *see* jasper
growing crystals, 148–51

healing, 18–19, 22–8
  absent, 27–8
  aura, 21–2
  contact, 24
  elixirs (symptoms), 135–47
  emotional, 26–7
  group, 27
  self, 25
  spiritual, 27
  stones, 29–35

iron pyrites, *fc*,82–3

jasper, 31, 132
  green, *bc*,31, 82, 132
  light green, *bc*,84
  pink and grey, *bc*,85
  red, *fc*,88–9
  variegated, *bc*,83

labradorite, *fc*,83
light green jasper, *see* jasper

meditation, 9, 61
  techniques, 15–17

mediumship, 59–63
moss agate, *see* agate

pendulum, 113–26
petrified wood, *fc*,84
pink and grey jasper, *see* jasper
programming, 15
psychic development, 49–52
  techniques, 52–8
purple agate, *see* agate

quartz, 33, 133
  agate, *bc*,78
  citrine, *bc*,30, 81
  crystal, *fc*,9, 37, 85, 113
  rose, *bc*,85–6
  rutilated, *bc*,86, 133

red jasper, *see* jasper
rose quartz, *see* quartz
ruby, *fc*,34, 86, 133
rutilated quartz, *see* quartz

serpentina, *fc*,34, 87

tektite, *fc*,87
tiger's eye, *fc*,34, 87–8, 134
turritella agate, *see* agate
turquoise, *bc*,35, 88, 135

variegated jasper, *see* jasper

white and amber agate, *see* agate

Of further interest

The Fortune-Teller's Workbook

A Practical Introduction to the World of Divination
by Sasha Fenton

Here at last is a book that takes the mystery out of divination — the DIY manual of fortune-telling.

The Fortune-Teller's Workbook contains sections on a wide variety of divination techniques, not only the well-known ones but also some of the lesser-known forms such as flowers and witchdoctor's bones. Many of the divinations show how to foretell the future, others explore the interesting and useful craft of character reading.

From a wealth of personal experience, together with advice and assistance from her many sources, best-selling author **Sasha Fenton** provides an enthralling insight into fortune-telling in all its varied and colourful forms. Among the range of fascinating and eye-opening subjects explored, the book examines fortune-telling by:

Tarot and crystal ball
*
numerology and runes
*
playing cards and pendulum
*
astrology and tea leaves
*
dominoes and palmistry
*
Chinese astrology and dice